D0374820

Gatherings of Angels

Gatherings

of Angels

Migrating Birds
and Their Ecology

EDITED BY KENNETH P. ABLE

With contributions by

Kenneth P. Able

James Baird

Keith L. Bildstein

William A. Calder

Sidney A. Gauthreaux Jr.

Brian A. Harrington

Gary L. Krapu

Frank R. Moore

Stanley E. Senner

Maps drawn by

Cindy Lippincott

COMSTOCK BOOKS

an imprint of Cornell University Press

Ithaca and London

First published 1999 by Cornell University Press

Printed in the United States of America

Library of Congress Cataloging-in-Publication Data
Gatherings of Angels : migrating birds and their
    ecology / edited by Kenneth P. Able ; with
    contributions by Kenneth P. Able . . . [et al.] ;
    maps drawn by Cindy Lippincott.
        p.   cm.
    Includes index.
    ISBN 0-8014-3362-2 (cloth : alk. paper).
    1. Birds—Migration. 2. Birds—Ecology.
    I. Able, Kenneth P.
QL698.9.G38   1999
598.156'8—dc21   98-47920

Cornell University Press strives to use environ-
mentally responsible suppliers and materials to
the fullest extent possible in the publishing of its
books. Such materials include vegetable-based,
low-VOC inks and acid-free papers that are recycled,
totally chlorine-free, or partly composed of non-
wood fibers. Books that bear the logo of the FSC
(Forest Stewardship Council) use paper taken from
forests that have been inspected and certified as
meeting the highest standards for environmental
and social responsibility. For further information,
visit our website at www.cornellpress.cornell.edu.

Cloth printing   10 9 8 7 6 5 4 3 2 1

Contents

Preface

In our enthusiasm to glorify and embrace nature, superlatives have lost their potency to describe its grandeur. We find ourselves repeating the same cliches. Those of us who write about bird migration are as guilty as anyone, but the facts of the matter are such that there is really no need for hyperbole. Twice each year, billions of birds, entire species, swarm across the globe, traveling thousands of miles as they follow the sun to populate regions that are habitable for only part of each year. The spatial scope of these migrations exceeds all other biological phenomena. So fantastic are they that ancient civilizations devised a host of myths to explain the periodic appearance and disappearance of such vast numbers of animals. Those apocryphal stories were concocted in part because what we now know to be true seemed then so completely beyond the pale. It seemed more likely that swallows buried themselves in the mud at the bottoms of ponds than that they flew all the way from Europe to Africa and back twice each year. But the truth turned out to be more amazing than the myth.

I have been awed by bird migration since I was a teenager. I became hooked the first time I saw through a telescope a night-migrating bird pass across the face of the full moon: a winged speck, so high that it seemed to take forever to cross the moon, apparently alone in its appointed mission. But although it may have been flying solo, it was not exactly alone. A few minutes' watching revealed others passing, silently and (so far as I could tell from the ground) undetected by almost all of humanity. The moon provided a tiny window into a secret and, for me, immensely exciting world. If so many birds could be seen passing across that half-degree dot that is the full moon, the sky must be filled with a vast swarm of birds.

Years later I watched those swarms flood a radar screen. Early radar operators had called the echoes from birds and other unknown targets "angels." Indeed, there is something almost miraculous about their explosive appearance just after dark as the woods, fields, marshes, and mud flats disgorge their avian contents into the night sky. The numbers are sometimes nearly incredible, hundreds of thousands of individual birds crossing a line one mile long every hour for most of the night. Seeing migration in action ignited a litany of questions. How did the birds know where to go? How would they ever be able to find their way there? How far would they fly before stopping? How do they know where to stop? How can their tiny and apparently frail bodies endure the rigors of this journey? These and other questions about how migration behavior works have fascinated me ever since, and studying them has occupied most of my professional career.

When I first became interested in migration, and even when I began to study it seriously, few people thought that the vast bird migration systems of the world could ever become, in the words of Lincoln Brower, "endangered phenomena." Even today, when one sees the skies over Nebraska blackened by sandhill cranes in the spring, when each summer the forests of the Northeast fill up with the songs of thrushes, warblers, and others, it is hard to imagine that there is any cause for concern. And yet we know that not so long ago the passenger pigeon was probably the most abundant bird in North America, Eskimo curlews darkened the skies, and Bachman's warblers were numerous enough to be shot to decorate ladies' hats.

The lifetime track of a migratory bird comprises a chain of events and places linked by the birds' own movements. The success of a migratory life history depends upon the strength of each link in that chain—a stopover place where a tired and hungry migrant can rest and refuel is just as necessary as a place to nest. Less attention has been paid to the biological problems faced by birds during the migratory journey, but the hazards of such a life on the road are legion. The terrain and habitats will often be unfamiliar. The potential foods may be new and unknown. It must be easy to become lost or to be displaced from the migration route by wind or storms. The energetic demands of long-distance flight are great and must be met efficiently in strange and often hostile places. Oceans, mountains, and deserts must be crossed, predators avoided. Twice each year, billions of birds run this gauntlet.

That many of them manage to do so successfully provides testimony to the extraordinary abilities with which evolution has equipped these global travelers. It is this critical part of their lives that we will explore in the following pages.

From the great spring flights of songbirds across the Gulf of Mexico to the massing of sandhill cranes on the Platte River, from the remarkable migration of the 3.5-g rufous hummingbird (0.12 oz; about the weight of a stick of chewing gum) to the vast intercontinental flights of shorebirds from their staging areas and the preparation of the blackpoll warblers for their endurance marathon, we will take you to places where migrants congregate—the gathering places of the radar angels. We will examine why migrants concentrate in these places and what they do there, why they take the routes that they do, and why they migrate when they do.

This book is intended for the layperson interested in birds or natural history who wants to learn more about the basic biology of bird migration. I have been most fortunate in being able to assemble in these pages contributions from some of the leading experts on bird migration in the Western Hemisphere. They can speak firsthand about the research they have done and what they have learned about the migrants with which they have spent so much of their lives. Because they have lived with the birds in these places, they are uniquely qualified to convey not only the biology but also some sense of the beauty and excitement that attend this most extraordinary of natural spectacles. I am most grateful to them for their enthusiastic participation in this project.

We begin with two chapters of general introduction about bird migration. The information contained therein should provide sufficient background to enable the reader to understand and interpret the material in the chapters that follow. If you are already familiar with the basics of bird migration, the first two chapters may be skipped. In the other chapters, the authors have provided thorough overviews of the research they have done on these migrants. They have tried to avoid technical jargon, but have not watered down the scientific issues. Because each of the chapters on individual migrant species has been written by a different author, there are inevitable differences in presentation and writing style. I have not attempted to rewrite those chapters to create homogeneity of style. Just as it is a privilege to hear about all of these studies of bird migration firsthand from the biologists who have done the work, I believe it is worthwhile to hear it in their own words and their own style. I hope you will agree that this adds to your enjoyment of the book and its value to you.

The idea for this book originated with Robb Reavill, formerly Science Editor with Cornell University Press. I am sure that the book has departed from her original conception in various ways, but I hope she likes the final product.

Acknowledgments

Peter Berthold, Gary Felton, and David Weintraub went out of their way to obtain and provide photographs for inclusion in the book. Mary Able read all of the chapters and provided useful comments and editorial assistance. Finally, I want to thank Peter Prescott, our editor, for the enthusiasm and professionalism with which he guided this project to completion.

Jim Baird (Chapter 5) is grateful to David B. Wingate for sharing data he has gathered on blackpoll warblers over his many years as the Conservation Officer of Bermuda, but most especially for allowing him to use critically important data he obtained on the Argus Tower off Bermuda in 1967.

Keith Bildstein's contribution (Chapter 6) would not have been possible without the long-term Hawk Mountain count efforts of Maurice Broun, Alex Nagy, Jim Brett, Laurie Goodrich, and many other Sanctuary staff and volunteers, who have helped create and maintain Hawk Mountain's long-term raptor migration database. Eric Atkinson helped calculate broad-wing energy expenditures, Wendy Scott helped prepare figures, and Nancy Keeler offered many useful suggestions on the manuscript.

Gary Krapu (Chapter 7) would like to thank Janet Keough and Robert Cox of the Northern Prairie Wildlife Research Center, Gary Lingle, Platte Watershed Program Coordinator for the Institute of Agriculture and Natural Resources at the University of Nebraska at Kearney, and David Carlson of the U.S. Fish and Wildlife Service Ecological Services Field Office, Grand Island, Nebraska, for their helpful comments on his manuscript.

Brian Harrington (Chapter 8) notes that much of the information about white-rumped sandpipers came from cooperators in the International Shorebird Surveys, from research of Raymond McNeil, Frans Leeuwenberg, Susana Lara Resende, and Paulo Antas. Logistic, financial, and administrative support was received from Manomet Center for Conservation Sciences, from the World Wildlife Fund-US, and from the U.S. Fish and Wildlife Service. He is grateful to the hundreds of other people who worked equally hard, but are unnamed here.

Bill Calder's long-term project on rufous hummingbirds (Chapter 10) has been supported by grants from the National Geographic Society and the National Science Foundation, by teaching positions at Rocky Mountain Biological Lab, and by Perky Pet Corp. which provided feeders. Lorene Calder helped unstintingly with the field work and provided perceptive second opinions that were unfettered by a biologist's notion of how things had to be (but sometimes weren't). The field work would not have been possible without the enthusiasm and hard work of many assistants. Nick Waser, David Inouye, Elly Jones, Kay Burk, Steve and Ruth Russell, Chris Otahal, Joan Day Martin, Sarahy Contreras, and Sarah Stapleton-Calder generously shared banding data. The staffs of Rocky Mountain Biological Lab and the Instituto Manantlan de Ecologia y Conservacion de la Biodiversidad at Las Joyas provided facilities and assistance.

KENNETH P. ABLE
Albany, New York

Gatherings of Angels

1

KENNETH P. ABLE

The Scope and Evolution of Bird Migration

What Is Migration?

Where I live in Upstate New York, winters are long and often harsh. In the depth of the frigid season I can expect to see ten or a dozen species of birds in my rural yard. Most of them are there because of the well-stocked feeders that we maintain for their benefit and our entertainment. Six months later, I could expect to encounter about four times that number of species. Except for the bats that live in our attic and disappear with the bulk of our birds, the mammal fauna on our property remains the same year round. Why is this?

Animals that live in environments that are strongly seasonal or for other reasons are habitable for only part of the year must evolve means of coping with the hostile period. Many species, including a few birds, escape by entering an inactive, dormant state (e.g., hibernation). This is the strategy adopted by some of our backyard mammals. Highly mobile animals have another option: to escape by leaving the area entirely and moving to a more hospitable region. Endowed with strong powers of flight, most birds have dealt with fluctuating environments by evolving migratory behavior [1].

What ornithologists think of as migration actually comprises a broad continuum of types of movement from sporadic irruptions of birds to the long-distance round-trips that we usually associate with the word. One can begin to understand the diversity of migration strategies by considering the ecological problem faced by the birds. Essentially, it is a problem of variability in resources. Be it food, water, cover, or competitors, the changing seasons can transform a comfortable environment into an unlivable one. The more variable the resources, the more extreme will be the coping solution. The other important factor about fluctuating resources is whether the variation is predictable. At high latitudes, it is certain that the abundant insect food of summer will be entirely absent in winter. Any bird dependent upon active insect life must always go elsewhere. In more temperate regions, things may not be so absolute. In some winters an absence of freezing periods might enable birds to overwinter successfully, whereas in other years a cold spell might eliminate the food supply and force a movement.

Different Types of Migration and the Environment

Two important factors need to be considered if one wishes to understand the spectrum of different kinds of movements that birds perform: the seasonal variability in the resources that birds need and the predictability of that variability.

In places where resources can be expected to be available year-round, there is no reason for individual birds to move long distances. Under those conditions, natural selection will favor individuals adopting *full-time residence*. By doing so, they avoid the risks and rigors of migration. Many familiar and common species such as most woodpeckers, chickadees, and titmice, the northern cardinal, and the northern mockingbird are resident species.

When the necessary resources in the breeding area fluctuate dramatically from season to season, and when this fluctuation is very predictable, *obligate annual migration* should evolve. This is the pattern seen in insect-feeders that nest in northern forests — many wood-warblers, vireos, thrushes, and flycatchers — all of which migrate southward for the winter.

The food of some birds fluctuates dramatically, but is more unpredictable and does not always follow a seasonal schedule. The seeds of coniferous trees and other seeds eaten by various finches such as crossbills, redpolls, pine siskins, and pine grosbeaks (Plate 1.1) are sometimes superabundant, sometimes entirely absent in a given region. The amount of food available varies not only seasonally but also unpredictably from year to year. It makes good sense for these species to have a very flexible migration strategy, one that is directly responsive to the availability of their special foods. Migrations by these birds are called *irruptive* because large numbers emigrate from the boreal forests in some years but remain at home in others. Similar irruptive movements occur in such tundra predators of rodents as snowy owls and rough-legged hawks.

Obligate migration and irruptive movements represent two extremes in the continuum of migration that is a response to the predictability and variability of resources. In between are intermediate situations that favor an intermediate solution, a type of behavior called *partial migration*. In partially migratory species, some individuals migrate whereas others are sedentary residents. Biologists envision two types of control mechanisms for partial migration. In one case, a population might be made up of two types of individuals, migrants and nonmigrants. The difference is presumed to be coded genetically: the migrant type would always migrate, the nonmigrant type would always behave as a resident. One would expect this solution to the problem to evolve when a) there are always sufficient resources to allow some, but not all, individuals to overwinter in the breeding area and b) when the degree of resource variability is very predictable from year to year. There is some evidence that

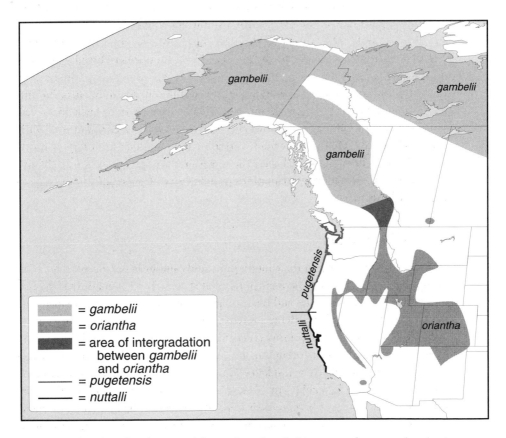

Map 1.1 The breeding distributions of the 4 subspecies of white-crowned sparrow that nest in western North America.

the European robin and southern European populations of the blackcap behave in this way [2].

In the absence of year-to-year constancy, we would expect a more flexible strategy to evolve in which the occurrence of migration and the number of individuals migrating vary from year to year in direct response to resource availability. In these cases, the individuals that migrate will not be predetermined genetically but may be predominantly the young or socially subordinate members of the population. The European blue tit seems to fit this pattern [3], and our North American chickadees may be doing the same thing during their occasional flight years.

As we have seen, what ornithologists call migration is actually a continuum of behavioral strategies that spans from permanent residency to obligate long-distance migration. Even with the numerous named variations on the theme described here, there are many cases that do not fit neatly into our preconceived categories. Even different populations of one species may exhibit quite different migratory tendencies. For example, Map 1.1 shows the distribution of the populations of white-crowned sparrows in western North America. The most northern breeding population, which is given the subspecific name *Zonotrichia leucophrys gambelii*, migrates to spend the winter on the southern Great Plains and into northern Mexico. Nesting farther south, the race *Z. l. pugetensis* migrates a short distance to the lowlands of central and southern California. The subspecies *Z. l. nuttalli* is a year-round resident in coastal California. During the winter, it is joined by migrants from farther north. The differences in migratory behavior found within the populations of this one species are obvious evolutionary adaptations to the degree of seasonality in the regions where the birds nest.

Migrants and Residents

In many breeding areas, on the wintering grounds and in between, migrants must occupy habitats that contain permanent resident species. It is axiomatic that the evolution of migratory strategies and the ecology of migrants will be intertwined with the biology of the residents. Being present in an area year-round gives the resident individuals an advantage in priority access to resources. Indeed, it has been hypothesized [4] that migrants can occupy an area only to the extent that surplus resources exist which are not utilized by residents. Migrants, then, are thought to be living off the fat of the land. The numbers of residents do not continue to grow until they consume all of the resources because their populations tend to be limited by mortality over the winter, when those resources are usually at their scarcest. In highly seasonal environments, then, there should always be an opportunity for migrants to exploit the flush of food during the temperate season.

There are obvious advantages and costs to both resident and migrant. Being a permanent resident frees a bird from the dangers and costs of migration and puts it on the breeding ground so that nesting can begin at the earliest opportunity. This is often an advantage because early nests tend to enjoy higher success. On the other hand, severe winters can kill many birds and this levy will fall disproportionately on residents. Migrants must pay the costs of migration, but they enjoy higher overwinter survival in the more moderate climates to which they move. To the extent that migrants and residents are competing for resources in the same locality, we might expect some relationship to exist between their populations. For example, when mild winters allow more residents than usual to survive and enter the breeding pool the next spring, there should be correspondingly reduced opportunity for migrants. Detailed studies in Britain have shown that the populations of resident species were indeed a function of overwinter survival, increasing in years with mild winters and decreasing in harsh ones [4]. Populations of migrants seemed to be controlled more by nesting success. When the populations of residents have been reduced by a severe winter, there should be increased opportunity the following spring for migrants that escaped the high mortality by spending the winter elsewhere. Conversely, when overwinter survival of residents is high, migrants might be faced with more stringent competition when they return in spring.

These kinds of interactions between individuals of populations that do not even occupy the same place all the time are extremely difficult to study directly. It is a subject that should receive more thought and attention. Should long-term global warming occur, resident species would likely be favored [5]. In settled areas of North America and Europe overwinter survival of resident birds may have been increased by bird feeders. Both factors could lead directly to increases in the populations of the direct competitors of migrants. If it seems far-fetched that feeding birds could have an impact of such magnitude, consider the story of the British blackcaps to be discussed later in this chapter.

The Evolution of Migration

Behavior evolves under the influence of natural selection just like any other trait of an animal or plant. In order for natural selection to work, there must be some raw material upon which it can act. That raw material is variation: individuals differ one from another with respect to some trait, be it wing length or the propensity to migrate. In order for evolution to occur, the variation must be heritable, that is it must have a genetic basis. It is obvious to any experienced bird-watcher that individuals of the same species differ in their migratory behavior. Birds migrate in different directions, for example. Some wind up in wildly unusual places where they generate much excitement among the birding fraternity, which refers to them as vagrants. Some individuals migrate farther than others, and some do not migrate at all and attempt to spend the winter within the breeding area. Observing this sort of behavioral variability is one thing; showing that it is under genetic control is quite another. It has long been assumed that many aspects of migratory behavior are inherited, but only recently have researchers revealed some of the details.

The blackcap (Plate 1.2) is a small warbler that breeds throughout western Europe from Scandinavia southward to areas around the Mediterranean Sea. Populations in the northern part of the breeding range are obligate long-distance migrants, those from around the Mediterranean are partial migrants, and those on the Canary and Cape Verde Islands are residents. This wide range in migratory habits within the same species makes the blackcap an ideal study organism, and Peter Berthold and his colleagues in Germany have taken advantage of this in their studies. Blackcaps from all of the populations can freely interbreed with one another, and this makes it possible to study in captivity the genetic control of various aspects of their migratory behavior.

A remarkable degree of genetic control over many facets of migratory behavior has been exposed through crossbreeding experiments. Nocturnal migrants such as the blackcap, when held in cages during the migration season, engage in a great deal of hopping and fluttering during the night. Outside the migration season, they sleep quietly on their perches. The amount of this migratory hopping (called *Zugunruhe*, literally "migratory unrest") seems to be related to how far a bird would migrate if free-flying. Long-distance migratory blackcaps show more hopping for a longer period of time than birds from short-distance populations. When long-distance migrants were bred with birds from the Canary Islands, which do not migrate, their offspring showed intermediate amounts of *Zugunruhe*.

Similarly, the direction of orientation during the first migration showed a high degree of heritability. Blackcaps from the western part of central Europe migrate southwestward and go around the western end of the Mediterranean; those from eastern Europe fly around its eastern end. Hybrid offspring of parents from the two sides of this "migratory divide" showed intermediate southward orientation [2].

These and other experiments performed by Berthold and his associates point to strong genetic control over many of the key components of migratory behavior. They suggest that in nature, if there existed potent selection by the environment, many aspects of migration might evolve quite rapidly. Again, the blackcap provides an illuminating example.

The blackcap is a common breeding bird in the British Isles, but until recent years did not overwinter there. Over the past 25 years or so, this pattern has changed. Increasing numbers of blackcaps began to spend the winter, especially in the more southerly parts of England and Ireland. This has become possible in large part because of the increase in feeding of wild birds in recent years. The really surprising thing about this evolutionary adventure is that the winter blackcaps are not birds that nest in Britain or Ireland; rather, they are blackcaps from central European populations that have migrated northwest in autumn instead of flying to ancestral African winter quarters. Because they have bird feeders to rely upon in winter and a shorter, lower-cost migration to Britain versus that to central Africa, the individuals who have taken up this novel migration pattern may enjoy higher overwinter survival. Birds wintering in Britain may also return to nesting places on the continent earlier in the spring and thus reap a reproductive advantage relative to their southbound cousins.

The key to any evolutionary novelty lies in the presence in the population of the appropriate genetic variation: in this case, birds predisposed to migrate northwestward in autumn instead of southwestward. When Andreas Helbig screened the fall orientation directions of German blackcaps, he did indeed find a few individuals that orient northwest. And birds trapped from the British wintering population, taken to Germany and tested there in the fall, also oriented toward the northwest. This variant migration direction breeds true and persists through two generations of offspring from crosses between birds that wintered in Britain. Before the advent of large-scale winter bird feeding, this genetic variant would have been selected against because conditions were unfavorable for a blackcap to survive winter on the British Isles. With that anthropogenic change in the environment, this minority migration direction became a viable option and the genes coding for it have presumably increased in the population.

On this side of the Atlantic, something similar seems to be happening with the rufous hummingbird, the featured player in Chapter 10. The migration pattern described by Bill Calder is typical of the species, and was the only pattern until very recently. Before the past decade, rufous hummingbirds were considered vagrants in most of the southeastern United States. In the past few years, however, their numbers have increased explosively until now many hundreds overwinter (mostly at hummingbird feeders) in the Gulf States. This striking phenomenon has not been studied in detail and it will be hard to do so, hummingbirds being more difficult than blackcaps to work with under laboratory conditions. But superficially the situations seem very similar: lost or misdirected individuals that formerly would have been weeded from the gene pool suddenly find themselves in a world made favorable by the hand of man. They provide daunting evidence of the dramatic and rapid impact that our own activities can have on a biological phenomenon so ancient as migration. Although the effects in these cases were unintentional and largely unforeseen, they perhaps provide a bit of hope that some migration systems may be able to adapt to the accelerating change that our own species is bringing to the earth.

The stunning thing about all of this is that it now appears that behavior as complicated as migration can evolve with remarkable rapidity. It has seemed apparent for a long time that migratory behavior appears and disappears repeatedly in the lineages of birds. In simplest terms, migration begins to evolve when individuals that move from one area to another rear more offspring than conspecifics that remain in one place. Changing environmental conditions provide the filter that determines whether migratory or sedentary habits are favored. Might it be possible to witness within a lifetime the evolution of migratory behavior in a sedentary bird?

In the early 1940s, house finches (Plate 1.3) from a nonmigratory population in California were released on Long Island, New York. The finches took hold and have since spread throughout the northern half of the eastern United States, becoming one of the most numerous birds in urban and suburban areas of the Northeast. Now established in this far more seasonally variable climate, migration has appeared [6]. Although the situation has not been studied in detail, it appears that the eastern house finch has become a partial migrant. Some individuals are resident year-round in the same area, while others regularly migrate back and forth to the Gulf States. This change has taken place in fewer than 50 years. It is now apparent that migratory behavior can sometimes appear anew or disappear from a population in just a few generations. It was surely some migrating dark-eyed juncos gone astray that colonized Isla Guadalupe, about 150 km off the coast of Baja California; and today the junco is established on Guadalupe as a sedentary population. Similarly, if less dramatically, populations of white-crowned and savannah sparrows along the coast of California have abandoned the migratory habit.

Controlling and Synchronizing the Annual Cycle

The annual cycle of a bird comprises a suite of integrated events of which migration is only one. Birds molt once or twice a year, come into and wane from breeding condition, and prepare for winter as well as migrate. In much the same way that the daily activities of all organisms are coordinated by biological clocks called circadian rhythms, the events of the annual cycle are regulated by a circannual clock with a much longer periodicity. As their name implies, these rhythms have a cycle length of about one year.

To discover whether animals have internal clocks of these sorts, biologists place birds in constant conditions in the laboratory and ask whether the rhythms persist. In the case of circannual rhythms, this requires keeping birds in captivity for many years under conditions that provide no information about the changing seasons outside (constant dim light or days of constant length). Birds kept in this way continue to go through cycles of migratory activity, breeding condition, molt, migratory activity, etc., in the same sequence as wild birds but gradually becoming out of synch with them. Such cycles have persisted for more than ten years in small songbirds that would not even live that long in nature [7]. So in obligate migrants, migratory behavior and its associated physiological changes are triggered internally by the bird's built-in clock, rather than by external conditions in the environment.

Outside the laboratory, of course, birds are exposed to many changes associated with the seasons. Foremost among these is the change in day length. It is that change which synchronizes the events of the annual cycle under natural conditions. Changes in the length of the daylight period, or *photoperiod*, have the power to entrain or set the circannual clock and keep it in phase with the natural world. Migratory condition is a syndrome that involves a host of behavioral and physiological changes. Among nocturnal migrants, birds that ordinarily sleep all night become restless during darkness. Many begin to consume more food, and some alter their diet and begin to eat things they do not eat at other times of the year. The metabolism of many species changes so that a greater proportion of the calories eaten is converted into subcutaneous fat deposits which serve as flight fuel. All of these events must be timed in a precisely coordinated way, and that is accomplished by circannual rhythms fine-tuned by changes in photoperiod (Plate 1.4).

It is obvious that when ornithologists speak of "migration," they are talking about a broad spectrum of behavioral solutions to the ecological problem posed by variability in the environment. Migration has evolved repeatedly within the lineages of birds, and its patterns are ever changing as natural selection molds populations in response to new or changing conditions. Across the spectrum from facultative movements to obligate migrations there is a parallel trend toward increasingly rigid internal control of behavior. The movements of facultative migrants are largely controlled by external conditions such as food availability, whereas obligate migration is triggered by internal physiological changes under the control of circannual rhythms, often before any deterioration in immediate environmental conditions is evident. It seems reasonable to suppose that the ability of a migratory species to survive large-scale environmental change may be in part a function of its position on this spectrum. Those with more rigidly programmed, internally controlled migratory habits may be less able to adapt to rapidly changing conditions and will be the most vulnerable if faced with drastic changes in climate, habitat, or resources.

The life of a migratory bird is played out on a global stage. The success of this evolutionary strategy depends on many variables—conditions on the breeding and wintering grounds as well as conditions along the path connecting those destinations. In many ways (energetically, in terms of risk, etc.) migration is the most demanding event on a bird's annual calendar and thus potentially the weak link in its life history. Coping on a day-to-day basis with the rigors of migrating long distances requires an elaborate suite of behavioral and physiological adaptations to which we will next turn our attention.

References

1. Dingle, H. 1996. Migration. The Biology of Life on the Move. Oxford: Oxford Univ. Pr.
   A very thorough and readable book-length account of all aspects of migration in all animals.
2. Berthold, P. 1996. Control of Bird Migration. London: Chapman & Hall.
   A technical account summarizing the literature on the control physiology of bird migration; especially useful in summarizing the voluminous studies of Berthold and his colleagues.
3. Smith, H.G., and J.A. Nilsson. 1987. Intraspecific variation in migratory pattern of a partial migrant, the blue tit (*Parus caeruleus*): an evaluation of different hypotheses. Auk 104:109–115.
   An original research paper describing the results of an intensive study of this European partial migrant.
4. O'Connor, R.J. 1990. Some ecological aspects of migrants and residents. In Bird Migration: Physiology and Ecophysiology (E. Gwinner, ed.). Berlin: Springer-Verlag.
   Reviews the relationships and interactions between populations of migrants and residents, especially in the British Isles.
5. Berthold, P. 1993. Bird Migration: A General Survey. Oxford: Oxford Univ. Pr.
   A general, though somewhat technical, introduction to all aspects of bird migration from methods of study to physiological control, evolution, navigation, and conservation.
6. Able, K.P., and J.R. Belthoff. 1998. Rapid 'evolution' of migratory behaviour in the introduced house finch of eastern North America. Proc. Royal Soc. Lond. B 265:2063–2071.
   An original research paper that examines the pattern of migration in this "new" migrant.
7. Gwinner, E. 1990. Circannual rhythms in bird migration: control of temporal patterns and interactions with photoperiod. In Bird Migration: Physiology and Ecophysiology (E. Gwinner, ed.). Berlin: Springer-Verlag.
   A thorough technical review of the control of the annual cycle in birds, with a particular focus on migration.

2

KENNETH P. ABLE

*How Birds
Migrate:
Flight Behavior,
Energetics,
and Navigation*

One swallow does not make a summer, but one skein of geese, cleaving the murk of a March thaw, is the spring. A cardinal, whistling spring to a thaw but later finding himself mistaken, can retrieve his error by resuming his winter silence. A chipmunk, emerging for a sunbath but finding a blizzard, has only to go back to bed. But a migrating goose, staking two hundred miles of black night on the chance of finding a hole in the lake, has no easy chance for retreat. His arrival carries the conviction of a prophet who has burned his bridges.
—Aldo Leopold, *A Sand County Almanac,* 1949

The northern coast of the Gulf of Mexico can be one of the premier places in North America to witness the drama of songbird migration, as Sid Gauthreaux and Frank Moore describe in their chapters. My experience there is much more limited, but has still left some vivid impressions. Nocturnal songbird migrants cannot make the entire trans-Gulf flight during the night, so most arrive during daylight hours. If you lie on your back and look skyward with a telescope or binoculars, you can sometimes see the arriving flight passing high overhead. They come in flocks, but they are nearly always too high for identification. Most continue inland without hesitation when the weather has been favorable during the water crossing. If you watch long enough, you may see a flock begin to hesitate and land in the coastal woodlands.

Once, many years ago, on Grand Isle, Louisiana, I watched such a flock begin to break up as individual birds circled briefly as if to make up their minds, then plummeted to earth. Their descent was so fast that it was impossible to follow with a scope. One instant they were specks near the limit of visibility in the sky, the next they vanished in a free-fall. And then they reappeared in the dead top of a live oak as if they had been beamed there—a chattering flock of eastern kingbirds, typically one of the earliest arrivals on a day of trans-Gulf migration.

They show no outward sign of having just traveled 500 miles or more nonstop under their own power. They do not look like marathon runners at the end of the race. In fact, they act like they might have been sitting there all morning. In just a few minutes they begin to sally forth and snatch insects from the warm spring air. This flock will probably spend the remainder of the day on Grand Isle, resting and feeding. Come dark, they will take off again on the next leg of their journey, as will nearly all of the other migrants that landed on the coast during the day. The next morning one can scour the woods in vain for passage migrants.

How does the behavior and physiology of a migratory bird support the incredibly strenuous activity of these flights? How do the migrants avoid becoming lost? How do they cope with the vagaries of weather? How do they find their way to precise destination points? (A longer but very readable account of these and other aspects of migration can be found in Paul Kerlinger's book [1].)

Physiology of Migration

Birds are, of course, the most aerial of all creatures, and flight has been a part of avian life for most of their evolutionary history. They are exquisitely adapted to life on the wing: extraordinarily efficient hemoglobin and a lung and air-sac respiratory system enable maximum oxygen uptake; a lightweight skeleton with hollow bones and internal organs of reduced size lighten the load. Even so, flight is strenuous activity and it takes a lot of calories of energy to propel a body through the air for hundreds of miles.

Fat stored in the body provides the fuel for long migratory flights. Gram for gram, fat is the most energy-rich substance that animals produce and store: when used, it yields about twice as many calories of energy as the same amount of carbohydrate or protein. Most of the fat stored prior to and during migration is deposited in *fat bodies* just under the skin. The most obvious of these lie over the abdomen and in the depression formed where the clavicles (wishbone) fuse. Blowing against the feathers of a long-distance migrant like a blackpoll warbler with a full fat load reveals a butterball nearly completely clothed in a layer of yellowish fat just beneath the skin. Ornithologists studying migration examine birds in this way routinely, using the amount of fat as an indicator of the bird's energetic condition and potential to continue migration right away.

As one might expect, long-distance migrants or those that must cross large expanses of water or inhospitable habitat lay down larger amounts of fat prior to embarking on migration. A typical nonmigratory bird usually carries 3% to 5% of its body mass as fat. For migratory songbirds, the figure is often 30% to 50%, or even more. Of course, starting out with such a heavy fuel load is a burden that costs energy and can reduce the maximum flight range of the bird, but the bird's load becomes steadily lighter as it flies and consumes more and more of the stored fuel [2].

Knowing both the amount of energy derived from burning fat as fuel and something about the flight speeds of birds, one can construct mathematical models to estimate just how far a migratory bird could fly on a given amount of fat. Different models use different assumptions and do not always agree in their predictions; however, they all agree that the possibilities are remarkable. Some shorebirds (Plate 2.1) should be able to cover about 6,000 miles (9,655 km) in a single flight, many songbirds can go 620 miles (about 1,000 km) without refueling, and even the 4.8-g rubythroated hummingbird has a single-flight range sufficient to carry it across the Gulf of Mexico [3].

The massive fat deposits associated with migration result from changes in eating habits and changes in metabolism. As the season of migration approaches, food intake in many species increases by as much as 25% to 30%. There may also be changes in diet, especially in preparation for fall migration when many species of songbirds, as well as some shorebirds, ducks, and gulls, dramatically increase the proportion of fruit in their diet. Fruit is high in carbohydrate, which is easily converted to fat by the body, and it is relatively easy to digest. As insect populations decline with the end of summer, many plants set fruit. That the production of fruit coincides with the migration of birds may not be entirely coincidental. The seeds of many plants are dispersed by birds, so what better time to produce seed-containing fruits?

Not only do birds eat more when they are in migratory condition, they select from the available smorgasbord those foods with the highest fat content. At the same time, their metabolism appears to change. The proportion of calories consumed that is converted to fat increases, and energy reserves already stored as carbohydrate may be converted over to fat [4].

The importance of fat and the ability to replenish fat reserves during stopovers on migration are themes that recur throughout the chapters that follow. Nothing is more critical to a migrant than its fuel dynamics. The arriving kingbirds I described had made the Gulf crossing in good weather and with a tail wind. They would have reached Louisiana with fuel to spare. That is why they can rest and feed for just a few hours and then take off again. Had they encountered bad weather out over the water, their situation could have been very different. When birds run out of fat to be burned as flight fuel, but cannot land, they begin to catabolize other body tissues such as muscle. This last-ditch strategy is surely successful in some cases in carrying the bird to a place where it can land. But when conditions are too bad, exhausted and emaciated birds cover offshore oil rigs and dead migrants wash up on the beaches. Such birds are not only devoid of fat, but their breast muscles (the ones used to power the wings) will also be eroded—they have literally burned out (see Chapter 4).

The strenuous exercise of flight generates heat, and this leads to evaporative water loss, mostly through breathing. Thus migrating birds are not only at risk of running out of fuel, but also of becoming dehydrated on long flights. During large fallouts of migrants on the northern Gulf Coast, I have often seen ten or more warblers of several species crowding around a puddle of water, drinking and bathing. It is easy to get the impression that they are exceptionally thirsty after their long flight, but field and laboratory studies suggest that fuel, not water, is usually the primary limiting factor in migratory flights. Heat and water stress, however, may have been factors favoring the evolution of nocturnal migration in many species.

Daily Timing of Migration

Birds crossing the Gulf of Mexico fly both night and day of necessity. But why do some kinds of birds migrate only at night when there is a choice, whereas others fly by day? In part, the difference is rooted in the requirements of different flight styles. Soaring birds like the broad-winged hawks that Keith Bildstein discusses (Chapter 6), as well as cranes and storks, literally use the energy of the sun to migrate. They depend upon rising columns of warm air (thermals) that result from the heating of the earth's surface by solar radiation. This kind of atmospheric structure, which the birds can use like an elevator to gain altitude with virtually no expenditure of their own energy, exists only during the daytime. As a result, soaring birds are constrained to the warmer portions of sunny days if they are to enjoy the maximal advantage of their body shape and soaring ability.

Some species, such as the white-rumped and western sandpipers featured herein (Chapters 8 and 9), other shorebirds and many waterfowl, migrate both during the day and at night (Fig. 2.1). For them, flying when the weather is most favorable for covering long distances seems to be more important than whether it is dark or light. Various kinds of smaller landbirds migrate largely or exclusively during the day. These include some woodpeckers, swallows, kingbirds, crows and jays, larks, pipits, bluebirds, American robins, blackbirds, and cardueline finches. The overwhelming

Fig 2.1 Shorebirds (such as these sanderlings) and waterfowl often migrate both day and night. (Photo by Frans Lanting/Vireo)

majority of passerine birds, however, migrate almost exclusively at night. Some may make low-altitude flights during the early-morning hours, a behavior that is especially noticeable in coastal areas but is not confined thereto (Chapter 5).

Ornithologists have speculated for decades over why so many normally diurnal birds migrate at night. Many ideas have been put forth [5]. Migrating at night frees up daylight hours during which the birds may feed and restore their fat reserves. The heating of the earth during the day, so important to soaring migrants, also produces a good deal of turbulence in the lower strata of the atmosphere. At night, the atmosphere is more stable and conducive to flight by slower-flying songbirds. The atmosphere at night is also usually cooler than during the day, and this may reduce the risks of overheating and dehydration during flight. As with so many "why" questions in biology, it is very difficult to obtain data that would allow one to discriminate among the several hypotheses. And although our minds like to focus on unitary explanations, there could be many factors simultaneously favoring the evolution of nocturnal migration. Remember, natural selection is opportunistic.

Most night migrants, if they are going to move on a given night, take flight within 30 to 45 minutes after dark. Their numbers in the sky increase rapidly, usually peaking before midnight, and then decrease steadily until dawn. If flying over land, most nocturnal migrants land long before daylight. Finding a proper place to put down in the darkness and in unfamiliar territory is perilous and sometimes leads to conspicuous disasters when waterbirds such as loons and grebes mistakenly land on wet roadways and parking lots.

Unless delayed by unfavorable weather, diurnal migrants usually initiate flight shortly after dawn. The number of migrants aloft reaches a maximum at around 10:00 a.m. and declines thereafter.

Alone or Together?

When we think of some kinds of migrants, we automatically conjure images of flocks. We don't envision a single sandhill crane winging across the sky; we see vast waving flocks of them spanning from horizon to horizon (Chapter 7). Geese in V-formation (Plate 2.2), kettles of broad-winged hawks, the wheeling masses of sandpipers over a mud flat, a long line of sea ducks skimming the dark waves—all familiar examples of migrants on the move. Night-migrating songbirds, however, do not fly in discrete flocks. Whether they are randomly dispersed in the airspace is not clear, but there is every reason to think that individual thrushes, warblers, and others are winging it alone up there in the darkness. The same birds, when forced to fly in daylight as in a crossing of the Gulf of Mexico, form up into flocks at dawn (see Chapter 3). As in the case of the kingbirds I watched land on Grand Isle, these flocks are often made up of a single species. How such segregated flocks form up from the mass of migrants aloft when darkness gives way to dawn is an intriguing question with no known answer.

Among large, long-lived species such as geese or cranes, family groups often migrate together. This provides at least an opportunity for young, inexperienced individuals to learn the migration route and traditional stopover sites from the old hands. For most songbirds, there is no evidence for such sharing of information. In many, adults and the young born that year migrate at different times during the fall season. The first-time migrants appear to make do with whatever innate information they possess and with what they can learn on their own as they make the journey for the first time.

How High to Fly

Before there were many real data on the matter, people believed that birds migrated at extraordinary altitudes. In some cases, they surely do: bar-headed geese regularly migrate over the highest Himalayas. Most do not, however, and with the invention of long-range surveillance radar during World War II it became possible to obtain accurate measurements on the height of bird migration. Most night-migrating songbirds are usually below 2,000 feet (600 m) when flying over land, and nearly all are lower than 6,500 feet (1,980 m). They will sometimes fly higher in order to reach altitudes with particularly favorable winds, occasionally reaching 15,000 feet (4,500 m) or more.

Shorebirds and waterfowl, larger and stronger of flight, often ascend to much higher altitudes. Even over land it is not unusual to find them at 15,000 to 20,000 feet (4,500–6,100 m). Flocks of shorebirds passing over the Antilles on their way to

South America (including, no doubt, many white-rumped sandpipers) have been recorded by radar as high as 22,000 feet (6,700 m) [6].

Diurnal landbird migrants usually fly at very low altitudes. Other daytime migrants such as soaring birds, waterfowl, shorebirds, and gulls may fly at considerable heights.

Flight altitude is strongly affected by weather conditions. Head winds and cloud cover generally have the effect of causing birds to fly lower than they would under more favorable conditions. Sometimes birds will ascend through clouds to reach the clear skies above, especially when winds are more favorable above the cloud layer. Whether and how the birds know that better flying will be found above the clouds remains unclear.

How Fast to Fly

The progress that a migrant makes during the course of its journey will depend on many things: its fat reserves and how quickly it is able to replenish them at stopover sites; how favorable is the weather for continued migration; its flight speed. Most passerine birds can achieve only relatively slow flight speeds in still air (20–30 mph). Waterfowl and shorebirds fly a good deal faster (30–50 mph). When organisms fly relatively slowly, the direction and speed of the wind can make a huge difference in their speed over the ground. A thrush flying with an air speed of 20 mph can achieve 40 mph over the ground if it flies with a 20-mph tail wind. On the other hand, a head wind of similar speed will stop it in place. When birds are embarking on a flight across water or a large desert, assistance from the wind may well mean the difference between life and death. It is thus not surprising that migrating birds pay strict attention to the weather. This theme recurs throughout the pages that follow, especially in Chapters 3 through 5.

When migrating over land, songbirds progress in a series of relatively short flights, up to 200 miles or so, interrupted by three to five days of rest. The length of stopover will, of course, depend on local weather, how much fat the bird has, and the refueling conditions. Larger, faster-flying waterfowl and shorebirds often make much longer nonstop flights. There are documented cases of ducks and geese flying up to 1,865 miles (3,000 km) from Canada to the Gulf Coast in just two days [3].

Weather and Migration

Once the bird is in migratory condition and on its way, weather is perhaps the single most important determinant of the migrant's day-to-day course. On a night with poor weather, the air can be virtually devoid of migrants. The next night, if condi-

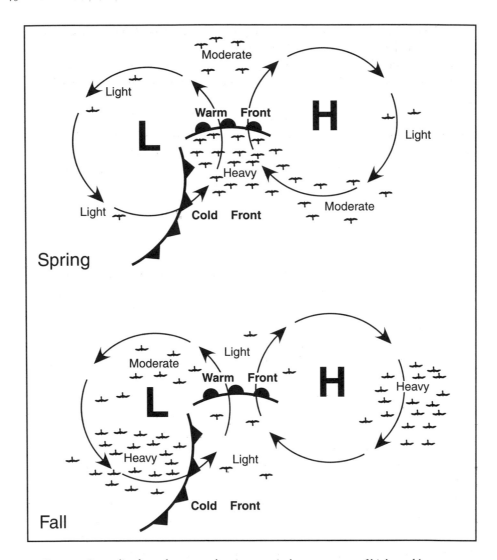

Map 2.1 Generalized weather maps showing a typical arrangement of high- and low-pressure centers with associated warm and cold fronts. In spring (*top*), migration will be heaviest in the warm, southerly winds on the back side of the high and ahead of the cold front. In fall (*bottom*), most birds will be migrating in the north or northwest winds and clearing skies behind the cold front. Reversed migrations (southward in spring, northward in fall) are almost always of much-reduced magnitude and occur in appropriate tail winds.

tions have become favorable, there can be tens, perhaps hundreds, of millions of birds aloft. When I was a graduate student studying migration with weather radar in northern Georgia, Sid Gauthreaux and I recorded a night in September when birds were passing at a rate of 200,000 individuals crossing a front one mile long every hour. This was not a local event, but rather a broad-front migration that went on for much of the night. The single most important variable in predicting when a watershed migration like this will happen is the weather.

What, specifically, about the weather do birds monitor? It appears that wind direction and change in temperature are the two most important factors. In autumn, migrants take off in the north winds and falling temperatures that prevail after the passage of a cold front. Northbound migrants in spring select the southerly winds and warming temperatures that are found on the western sides of high-pressure systems (Map 2.1). Migrants tend to avoid flying in rain, fog, low overcast and strong winds.

Although migratory birds seem to be quite adept "weatherpersons," extreme weather conditions can pose a significant hazard. Hurricanes and severe cyclonic storms are known to transport birds hundreds of miles out of range. Even under less-extreme conditions, migrants, especially slower-flying songbirds, may be blown off-course by crosswinds. In fall along the New England coast, large numbers of night migrants pass offshore when a cold front and its attendant northwest winds move out to sea. Many of the birds certainly perish at sea, but others land on ships and offshore islands such as Block Island, Rhode Island. Many of these birds can be seen embarking on northwestward flights back to the mainland during daylight.

Finding the Way

Ornithology is a science to which amateurs continue to make significant contributions. Among the most dedicated are the hundreds of bird banders who spend countless hours trapping and marking birds, maintaining accurate records, and carefully documenting and recording observations of previously banded individuals. With respect to migration, the single most important fact revealed as a result of all this effort is that many (perhaps most) migratory birds have the uncanny ability to return to previously occupied places, even though their intervening travels may have taken them thousands of miles away. How they accomplish these feats of homing is the central question that occupies the attention of those who study animal navigation. Much research has been done over the past thirty years, and the field remains very active; here I can provide only a thumbnail sketch of what we know about the process.

The first migration of a young bird is controlled by the endogenous circannual rhythms described in the preceding chapter. These youngsters seem to come into the world with some innate knowledge about the direction and approximate distance to fly on that first migration. With a good measure of luck, this information may be sufficient to get most birds within the winter range of their population, but it must be a rather imprecise business [7]. It seems quite clear that the first-time migrant has no knowledge of the specific goal of its journey. Once the bird has found an appropriate place to spend the winter, however, it will imprint upon that place. Thereafter, it will be able to navigate to this locality from almost anywhere. Some of the most conspicuous examples of this phenomenon are provided by vagrant birds that spend the winter in places far outside their normal range. Most such birds are immatures on their first migration, having gone astray because of faulty navigational equipment, inexperience, the vagaries of weather, or some other eventuality. Those that survive the winter and the following breeding season are very likely to return to their errant winter quarters as adults in subsequent years. Birders have learned to expect this and to watch for the vagrants' return.

On the first migration, then, birds know the appropriate direction and approximate distance to travel, but know no specific goal. To be able to home to a precise location requires direct experience with that place. Exactly what the requisite experience may be we do not know, but it is necessary for the bird to spend some time moving around freely in the "target" area, be that the natal region or the wintering ground. Once this reconnoitering has taken place, the bird will be able to return to the familiar locality from great distances and from places far beyond its region of direct familiarity. Homing pigeons are famous for their ability to return to their lofts from very great distances and, as such, they have been the "laboratory rat" of bird navigation studies. But extraordinary homing feats have been documented from many species of wild birds, both migratory and nonmigratory. The classic examples are from strong-flying seabirds such as Manx shearwaters and Laysan albatrosses, but more prosaic species such as the white-crowned sparrow have performed feats of similar magnitude.

Homing pigeons head off toward home within a few seconds of release. Is this exceptional, or do other birds behave similarly? Several years ago, my students and I spent a summer following wood thrushes as they attempted to return home after we had displaced them from their nests. Each thrush was equipped with a small radio transmitter (Plate 2.3) so that we could map its movements in detail. As with homing pigeons, these thrushes headed back toward their nests from the very beginning of their trip. There was no wandering about or searching; they knew where they were going from the outset. How might birds accomplish such homing feats?

Some very well controlled studies with homing pigeons, done in Germany, showed that the birds could home from unfamiliar sites hundreds of miles from the loft even if they had been prevented from perceiving any potential navigational information while being transported to the release site [8]. Other experiments have shown that pigeons will use information picked up during the outward journey from the loft, but it is not necessary for them to do so. They can figure out the direction toward home entirely on the basis of information perceived at the release site. The best explanation for this ability comes from what is called the map-and-compass model of homing navigation.

To understand how the map-and-compass process works, put yourself in the following situation. You have been blindfolded and driven, by a very circuitous route, to the middle of an unfamiliar forest. The blindfold is removed, you are handed a compass, and told to find your way home. In essence, you have been turned into a homing pigeon or one of my kidnapped wood thrushes. With your compass you can tell directions: for example, which way is south. But if you don't know which way is out of the forest or, more important, which way is home, the compass will be of little use. To make effective use of the compass you need to know where you are relative to your goal. In short, you need a map. Once the map tells you that you are south of home, for example, you can then employ the compass to identify north, the direction you need to begin walking. Virtually all of the data from bird-homing experiments are consistent with the notion that true navigation is a two-step process of this sort. First, some kind of global-positioning system or map is employed to identify spatial position; then a compass is used to identify the direction of movement indicated by the map. Much more is known about the compasses that birds use than about the nature of the map, so we will take a look first at the compasses.

Bird Compasses

When ornithologists began to search in earnest for the compass that birds use in homing navigation and migration, everyone assumed that we were looking for a single entity. What a surfeit of riches this quest has uncovered! We now know that birds possess a host of compasses that are related to one another in rather complicated ways, some based on information perceived with the eyes and some relying on senses whose very existence was doubted until recently [9].

The first bird compass to be discovered was one based on the sun's position in the sky. Birds, and many other animals, use the sun to identify directions in much the same way that we do. In order to use the sun as a compass you must know something of its path across the sky: that it rises in the eastern sky, passes through south at noon (if you are in the Northern Hemisphere), and sets somewhere in the west. You also need to know the time, and all animals are equipped with an internal clock synchronized to the 24-hour day. Homing pigeons use the sun as their compass of first choice when they find themselves released far from their loft, but how important the sun compass is to migrants remains something of a mystery.

Because most birds migrate at night, after the discovery of the sun compass it seemed logical to ask whether nocturnal migrants might use the stars as a compass. After all, we do so, but it takes some fairly sophisticated machines to achieve much accuracy. Early experiments performed in a planetarium where a replica of the starry sky could be manipulated showed that, indeed, birds do use the stars to take compass directions. Unlike the sun, the stars in the sky are myriad and the constellations that are visible change with time of night and season in a complicated way. Early notions of how the star compass worked supposed that birds were born with a genetically programmed star map in their brains. We know that some astonishing things seem to be subject to genetic coding (recall the time, distance, and direction program discussed earlier in this chapter), but an innate star map stretches credibility even today.

What seems to happen instead is that the birds learn the layout of the constellations by studying the apparent rotation of stars during the first summer of their lives. In this way, they locate the North Star (Polaris), that beacon of such importance to our own navigation, the one star that does not appear to move. Some inborn rule seems to tell the birds that this star is toward true north, or at least that they need to fly more or less away from this star if their program tells them to fly southward. The spatial relationships among the constellations are then memorized so that when it comes time to migrate, the birds no longer have to consult stellar rotation.

Two compasses down, a couple more to go. One of the most surprising discoveries in behavioral biology during the half century just past was that many animals possess a magnetic sense of direction. This remarkable capacity was first reported in migratory birds, and although it is still not clear where the sense organ is that perceives the earth's magnetic field, the existence of the magnetic compass has now been demonstrated in many species of migratory songbirds and the homing pigeon as well as in all the other classes of vertebrate animals [10].

The magnetic compass of birds, unlike our hand-held compasses, does not detect the polarity of the magnetic field. It is as if the needle on your backpacking compass were identical on both ends: you could see the north–south axis, but couldn't tell which of the two directions was north. Presented with this situation, the birds use an additional piece of information from the magnetic field. The lines of force that make up the magnetic field have an inclination, or *dip angle*, with respect to the earth's surface. In the Northern Hemisphere, for example, the field lines incline downward toward magnetic north. Birds use this clue to distinguish the direction toward the equator from that toward the pole. The initial discoveries and many of the details of how the magnetic compass works have been made in the laboratory of Wolfgang and Roswitha Wiltschko in Frankfurt, Germany. With birds in circular orientation cages in which night migrants hop repeatedly in the migratory direction, it is possible to use Helmholtz and other types of coils to manipulate the characteristics of the earth's magnetic field (Plate 2.4). By so doing, the orientation behavior of the birds can be altered in ways that reveal the existence and mode of operation of the magnetic compass [11].

Most nocturnal migrants initiate their journeys shortly after sunset, so it seemed reasonable that they might make important decisions about which way to fly at that time. Frank Moore and my students and I did many experiments over the past 25 years that showed this was indeed the case. In fact, visual information at sunset turned out to be of often overriding importance in the pantheon of orientation capabilities the birds have at their disposal [12]. Surprisingly, however, the visual cue used is not the sun itself, nor the bright glow in the western sky where the sun has dropped below the horizon. Rather, the birds look at patterns of polarized light in the sky. The earth's atmosphere acts like a weak polarizing filter, and sunlight that passes through that atmosphere produces a pattern of polarized light that is visible in the clear sky. The patterns reveal the position of the sun even if its disk is hidden by clouds (inasmuch as it is the sun's light that is being polarized) and, more important, they can be used to identify true north.

It has been known for decades that insects use polarized skylight as a surrogate for the sun, thus enabling them to perform solar compass orientation even when the sun is covered by clouds. By testing birds in orientation cages covered with large sheets of polarizing material between sunset and the time the first stars made their appearance, I was able to show that the birds were very responsive to manipulations of the polarized light. By changing its axis I could predictably change their orientation direction; and by removing the stimulus, by replacing the polarizing material with depolarizing sheets that eliminate the pattern, I could render them unable to select a direction. Unlike honeybees, the birds seem not to use polarized skylight simply as a means of figuring out where the sun is when it is not directly visible. Rather, the polarized light patterns provide a separate compass that is probably important especially at dawn and dusk, when those patterns are particularly apparent in the sky directly overhead.

Although humans cannot sense magnetic fields, we can consciously see polarized light from the sky even though most people are completely unaware of this obscure ability. We see polarized skylight as a small, bow-tie–shaped image long ago named Haidinger's brush. How birds see polarized light is not known and even whether they do is controversial in some quarters. By their behavior, however, the birds seem to say clearly that they can.

Why birds have so many different ways of determining compass directions is an interesting question, but one that can be answered only by conjecture. Presumably it is advantageous for migrants to have backup systems that they can call upon when they encounter the inevitable problems of migration. Clouds often obscure polarized light patterns and stars, but the magnetic field is ubiquitous. On the other hand, there are places where the geomagnetic field is severely disturbed. For species that migrate far south, the familiar stars of the northern sky will disappear below the horizon as unfamiliar ones appear to the south. All of these sources of variability should place a premium on having at hand several capabilities that can be consulted separately or in combination.

Navigational Maps

We know that many migrants have the ability to return to very precise locations, the ability to home. We also know that a compass alone is not sufficient to perform this task. A map or global-positioning system of some kind is also required. What kind of information is out there in the world that could provide a bird with the equivalent of its latitude and longitude? In theory, there are several possibilities. We use various systems based on celestial bodies, but to do so requires quite precise measurements and elaborate calculations. There is a good deal of evidence from homing pigeons that birds do *not* get their position in this way. Parameters of the earth's magnetic field could be used, and here the evidence is more equivocal. Homing pigeons respond to very small disturbances in the field: *magnetic anomalies* (places where the earth's field is locally distorted, usually as a result of large iron deposits near the surface) and so-called *magnetic storms* induced by sunspot and solar flare activity. Neither of these should have any effect upon a magnetic compass and yet they disturb the pigeons. Perhaps they influence the map. Recent experiments on hatchling sea turtles support the notion of a magnetic positioning system.

Far more surprising is the idea, put forth by Floriano Papi and his colleagues at the University of Pisa, that airborne odors form the physical basis of the homing pigeon's map. Surprising because conventional wisdom has held that birds in general have a very poor sense of smell. The hypothesis is that pigeons build up an odor map of their surroundings by learning that certain odors are associated with winds blowing from particular directions. Much like a local topographic map, the familiar region may be gradually extended through exploratory flights. This idea remains somewhat controversial despite an enormous volume of experimental evidence, most of which supports the hypothesis [13]. The overwhelming majority of the work has been performed on homing pigeons because they are much easier to work with than migratory birds. However, some homing experiments on European starlings and swifts suggest that they may employ the sense of smell in the same way. Is it possible that long-distance migrants could rely on an olfactory map also?

We know that many migratory birds show remarkable site fidelity with respect to breeding and overwintering locations. They home, sometimes after having traveled halfway around the world or farther. Ornithologists think that these birds must be doing something analogous to what a homing pigeon does (and thus we continue to use pigeons as a model system for study), but we know next to nothing about when and how this navigation takes place. Do the birds navigate throughout the journey, constantly adjusting for displacements from the correct course induced by wind or other factors? Or do they simply fly in the appropriate compass direction and perform the more refined navigation to the goal only during the final stages of the flight? Many flocking migrants such as shorebirds, waterfowl, and cranes seem to show great fidelity to particular migration routes and stopover sites. This fidelity suggests that perhaps they are navigating rather precisely throughout the journey. The occasional but very rare recovery of a banded songbird at the same migration stopover in different years makes one wonder whether they too might have such an extraordinary ability (see Chapter 10). But the truth is, we don't yet know.

What we do know is that despite nearly 50 years of intensive study punctuated by many startling discoveries, we still cannot explain in a detailed, mechanistic way how birds do what we know with certainty that they do: return with incredible precision of timing to pinpoint locations on the earth after traveling thousands of miles in the interim.

References

1. Kerlinger, P. 1995. How Birds Migrate. Mechanicsburg, PA: Stackpole Books.
   A popular account of all aspects of bird migration.
2. Blem, C.R. 1990. Avian energy storage. *In* Current Ornithology (D.M. Power, ed.). New York: Plenum.
   A thorough technical review of fattening and the energetics of migration.
3. Berthold, P. 1996. Control of Bird Migration. London: Chapman & Hall.
4. Bairlein, F. 1990. Nutrition and food selection in migratory birds. *In* Bird Migration: Physiology and Ecophysiology (E. Gwinner, ed.). Berlin: Springer-Verlag.
   Reports original research and reviews the literature concerning dietary aspects of the preparation for migration.
5. Kerlinger, P., and F.R. Moore. 1989. Atmospheric structure and avian migration. *In* Current Ornithology (D.M. Power, ed.). New York: Plenum.
   Critically reviews all of the hypotheses put forth to account for the fact that most birds migrate at night.

6. Williams, T.C., J.M. Williams, L.C. Ireland, and J.M Teal. 1977. Autumnal bird migration over the western North Atlantic Ocean. Am. Birds 31:251–267.

   Describes radar studies of the mass overwater migration from northern North America across the open Atlantic to South America.

7. Gwinner, E. 1996. Circadian and circannual programmes in avian migration. J. Exper. Biol. 199:39–48.

   Describes original research and reviews the interaction between internal clocks and environmental conditions encountered by migrants.

8. Wallraff, H.G. 1980. Does pigeon homing depend on stimuli perceived during displacement? I. Experiments in Germany. J. Comp. Physiol. A139:193–201.

   A very technical paper which describes some of the best experiments demonstrating that pigeons can determine where they are by using only that information which is available to them at the point of release.

9. Able, K.P. 1995. Orientation and navigation: a perspective on fifty years of research. Condor 97:592–604.

   A historical perspective on the development of our current understanding of how birds find their way.

10. Wiltschko, R., and W. Wiltschko. 1995. Magnetic Orientation in Animals. Berlin: Springer-Verlag.

    A thorough technical account of the occurrence, functions, and mechanism of magnetic orientation in all animals.

11. Able, K.P. 1998. A sense of magnetism. Birding 30:314–321.

    An account of magnetic orientation in birds written for the layman.

12. Able, K.P. 1993. Orientation cues used by migratory birds: a review of cue-conflict experiments. Trends Ecol. Evol. 8:367–371.

    A synthesis of how the different compass-orientation capabilities of migratory birds relate to one another.

13. Able, K.P. 1996. The debate over olfactory navigation by homing pigeons. J. Exper. Biol. 199:121–124.

    A review of and commentary on the long-standing controversy over whether pigeons use odors as a major part of their navigational map.

3

SIDNEY A. GAUTHREAUX JR.

Neotropical
Migrants
and the Gulf
of Mexico:
The View
from Aloft

Every yeere there passe
from the end of Cuba
infinite numbers of divers
sorts of Birds, which
come from the north of
the firme Land, and
crosse over the Alacrain
Islands and Cuba, and
flye over the Gulfe South-
wards. I have seene them
passe over Darien and
Nombre de dios and
Panama in divers yeeres
. . . so many that they
cover the skie: and this
passage . . . continueth a
moneth or more about the
moneth of March.
—Gonzalo de Oviedo,
Historia general y natural
de las Indies, 1535

The spring flights across the Gulf of Mexico are one of the great migration spectacles of the Americas. The groundings, or "fallouts," of migrants that can sometimes be seen at places such as High Island, Texas, the
oak cheniers of coastal Louisiana and Dauphin Island, Alabama, have become legendary among birders. With books and articles devoted specifically
to providing information about where, when, and how to witness such a fallout, it is hard to believe that whether songbirds even cross the Gulf in large
numbers was a controversial idea not so long ago. Although the evidence
marshaled by George H. Lowery, Robert J. Newman, and their colleagues
convinced most people that massive trans-Gulf migrations were a reality, the
case was largely circumstantial. All that changed when Sid Gauthreaux began
to observe migration in southern Louisiana with weather radar. Sid grew up

in New Orleans and there developed a fascination for migration through the region that turned into a life's work. As a teenager he witnessed some of the grandest groundings of trans-Gulf migrants anyone has ever seen. But on the screens of the radars ringing the northern Gulf Coast it was suddenly possible to see, in a single regional panorama, what could previously only be inferred from a glimpse here and a peek there. The scope and magnitude of trans-Gulf migration was revealed in a flash, and the way opened to study many aspects of the phenomenon. Sid Gauthreaux was the pioneer in these studies, and 30 years later he is still at it, with new questions, new techniques, and new insights. There is no one better qualified to bring us an overview of trans-Gulf migration from aloft. —*K.P.A.*

On their spectacular migrations to and from the breeding grounds, many birds cross large geographical areas that prevent landing (e.g., large bodies of water) or areas where food and water are scarce (e.g., deserts). The Gulf of Mexico is a formidable water barrier positioned between the breeding and wintering grounds of many migratory species that breed in central and eastern North America. Whether songbirds cross the Gulf of Mexico in spring and fall migration has been debated heatedly since the turn of the century, and as a result trans-Gulf migration has been thoroughly studied by ornithologists. The lessons learned from these investigations illustrate how science works and how sometimes both sides of a debate can be correct!

The Trans-Gulf Migration Controversy: The Turbulent 1940s and 1950s

In the late 1800s and early into this century Wells W. Cooke, an assistant biologist working for the Bureau of Biological Survey in the U.S. Department of Agriculture, analyzed the data gathered by numerous amateur and professional biologists throughout the country and discussed the patterns of bird migration in North America. At the time, the data were sparse and largely anecdotal, but Cooke thought that the distribution and migration patterns of many species were nonetheless clear. He was convinced that many different types of migratory birds flew across the Gulf of Mexico on their way to northern breeding grounds in spring and also crossed the Gulf in the fall when they returned to overwintering areas in the tropics. Moreover, he suggested that in spring the lack of some migrants (e.g., yellow-breasted chat, American redstart, and indigo bunting) along the coast of northwestern Florida was attributable to the fact that the migrants continued to fly inland after crossing the Gulf and flew over the observer (Plate 3.1). Cooke comments, "It would thus seem that the popular idea that birds find the ocean flight excessively wearisome, and that after laboring with tired pinions across the seemingly endless wastes they sink exhausted on reaching terra firma, is not in accordance with the facts" [1]. He further suggested that "endowed by nature with wonderful powers of aerial locomotion, under normal conditions many birds not only cross the Gulf of Mexico at its widest point, but may even pass without pause over the low, swampy coastal plain to the higher territory beyond." The hazards of migrating over large bodies of water did not go unnoticed by Cooke. He reported a catastrophe witnessed from the deck of a vessel 30 miles off the mouth of the Mississippi River during the spring, when a powerful cold front penetrated the northern Gulf and intercepted large numbers of migrating birds, mostly warblers, that had completed nine-tenths of their trans-Gulf journey. Hundreds fell into the Gulf and drowned. Cooke noted that such accidents are not infrequent.

In January 1945, George G. Williams, a professor at Rice Institute in Houston, Texas, challenged the traditional belief that many migratory birds take off from the Yucatán and the Campeche Bay region during the evening in favorable weather and fly northward directly across the Gulf of Mexico, reaching the northern Gulf Coast of the United States the next morning. Williams pointed out that the data in support of trans-Gulf migration were meager at best and largely consisted of the reports in Cooke's publications four decades earlier. Cooke's evidence in support of the theory was of three kinds: 1) some direct visual observations of migrating birds crossing the Gulf; 2) the fact that some migrant species were absent or extremely uncommon on the Texas and Florida coasts in spring but very abundant in the lower Mississippi Valley and along the Louisiana coast; and 3) that simultaneous arrival dates of many species along the entire northern coast of the Gulf suggested an overnight, trans-Gulf flight. Williams carefully examined the evidence and came to a very different conclusion [2]. He suggested the likelihood that the birds observed over the Gulf were coastally migrating birds that were displaced southward out over the Gulf by northerly winds and stormy weather associated with the passage of cold fronts in this area in spring. He refuted Cooke's analyses of the distribution of migrants along the Texas coast, the northern Gulf Coast, and the Florida coast and Keys and concluded that "no species appears in Louisiana and northwestern Florida that does not appear in at least equal numbers along the whole Texas coast, or on the keys and peninsula of Florida." The only logical interpretation, according to Williams, is that birds migrate around the Gulf and not across it (Map 3.1). With regard to the simultaneous arrival dates of many species along the Gulf Coast, Williams asserted that "the fact that spring arrival dates for any species may be very close together for the entire Gulf Coast does not prove that the species in question has migrated across the Gulf."

Five months after Williams published his paper, George H. Lowery Jr., a profes-

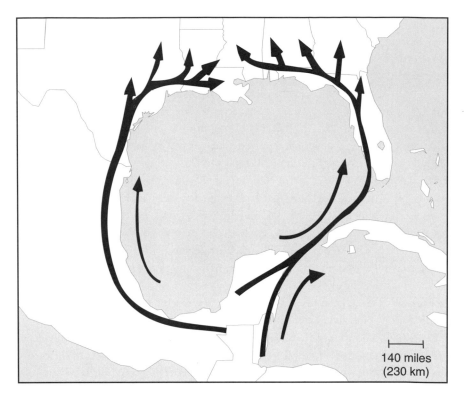

140 miles
(230 km)

Map 3.1 The Gulf of Mexico and surrounding regions showing the circum-Gulf migration routes discussed by George G. Williams in 1945.

sor at Louisiana State University, published in a different ornithological journal a paper on the trans-Gulf migration of birds and the coastal hiatus [3]. Lowery was unaware of the Williams paper and presented considerable data in support of the original position of Cooke on trans-Gulf migration and the paucity of trans-Gulf migrants near the coast on many days in spring. Lowery defined the *coastal hiatus* as the zone that trans-Gulf migrants fly over after arriving on the northern Gulf Coast. In favorable flying conditions, arriving trans-Gulf migrants might continue inland for considerable distances before landing; or, after a cold front or stormy weather, arriving trans-Gulf migrants might be "downed" in the first woodlands along the coast and the birds could subsequently take off and fly over much of the Coastal Plain to locations a couple hundred miles farther inland.

A year later Lowery addressed the points made by Williams in 1945, carefully re-examining the evidence Williams used in support of the notion that all migration northward in spring from the tropics was circum-Gulf rather than trans-Gulf [4]. In addition Lowery presented important new data in support of the existence of spring trans-Gulf migration. He used a moon-watching technique (Fig. 3.1) that enabled him to see the silhouettes of migrating birds as they passed before the disk of the full moon. By moon-watching at Progreso, Yucatán, in the spring of 1945, he was able to record migrating birds heading northward out over the Gulf of Mexico. In the eyes of the ornithological community, Lowery's second paper firmly reestablished the conviction that many and perhaps most of the migrants that breed in the eastern two-thirds of North America and winter in Central and South America reach their destinations in spring after flying directly across the Gulf of Mexico.

The scientific debate over the existence of trans-Gulf migration continued when Williams published a paper [5] that was solely devoted to a criticism of Lowery [4].

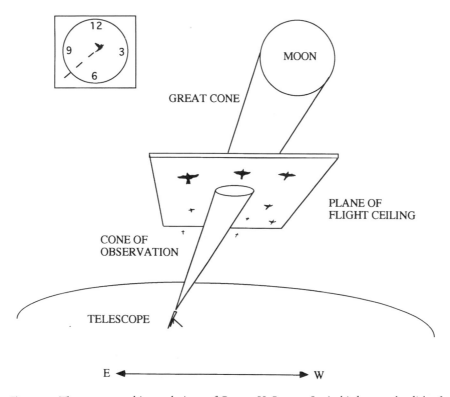

Figure 3.1 The moon-watching technique of George H. Lowery Jr. As birds cross the disk of the moon their flight paths are coded as "in" and "out" times on an imaginary clockface. All paths are then analyzed to produce a migration traffic rate—the number of birds crossing 1.6 km (1 statute mil of front) per hour.

He pointed out that Lowery's paper contained a wealth of new and valuable material but also a wealth of errors including lengthy straw-man arguments, misinterpretation of other observers' data, of original data, and of Williams's first article, cardinal omissions, and a tendency to formulate large general laws on the basis of minute fact. Not convinced by Lowery's criticisms and additional data, Williams reaffirmed his position that "vast numbers of birds regularly migrate around the sides of the Gulf of Mexico in spring; there is no valid evidence to show that any large number of birds regularly migrate across the Gulf in spring." Three years later [6], Williams confessed to some errors in his earlier two papers but still concluded that the coastal hiatus was "a lacuna south of and between two great spring migration triangles, one extending north and northeast from southern Texas, the other extending northwest, north, and northeast from Florida." He still believed that "periodic cold fronts, with northerly winds, striking the northern sides of these migration triangles, push migrants down against the coast, where they are often seen in great numbers immediately after the passage of a cold front." He further allowed that, "sometimes, the cold fronts push birds over the Gulf itself, where they have been mistaken for trans-Gulf migrants."

In his dissertation on the use of moon-watching to study quantitatively the nocturnal migration of birds, Lowery included a review of the trans-Gulf versus circum-Gulf controversy [p. 438–450 in Ref. 7]. There Lowery states clearly that he never believed that *only* trans-Gulf migration occurred in spring, and he acknowledges that he had long held that considerable circum-Gulf migration through eastern Mexico and southern Texas is a common occurrence in spring. He discusses moon-watching data that he and his colleagues had gathered from Tampico and Progreso, Mexico, in the spring of 1948 and concludes that "no other single piece of evidence so conclusively demonstrates that birds cross the Gulf of Mexico in spring in considerable numbers as do flight density data recorded from Progreso in 1948." As of 1954, 73 species of nonpelagic birds had been observed over open waters of the Gulf of Mexico, and many of these were small songbirds engaged in what was suspected to be spring trans-Gulf migration [8].

In an attempt to determine which of approximately 200 species use the circum-Gulf route, the trans-Gulf route, or both in their spring migrations to breeding grounds, Henry Stevenson, a professor in the Department of Biological Sciences at Florida State University, examined evidence gathered by cooperators between 1946 and 1955 from three sources: 1) direct observation of migrating birds; 2) the comparative abundance of birds around the Gulf in spring; and 3) the sequence of spring arrival dates [9]. The data indicated that both circum- and trans-Gulf routes are followed commonly in spring. Approximately 40 species of transients, summer residents, and winter residents were more common on the northern Gulf Coast than on its eastern and western sides. Stevenson pointed out that the larger numbers of species and individuals on the Texas coast were the result of both circum-Gulf and trans-Gulf migration routes, and he suggested that in Texas and Florida some of the northward circum-Gulf movements may actually have started as trans-Gulf flights that were displaced westward to the Texas coast or eastward to the Florida coast by strong winds over the Gulf. After landing, the migrants resumed their northward migration along the coast.

The Aftermath of the Controversy

The debates that took place in the scientific journals during the late 1940s and early 1950s stimulated considerable subsequent study of the patterns of bird migration in relation to the Gulf of Mexico. Much of what we know today about bird migration in the region was an outgrowth of the controversy. Lowery's moon-watching technique has been used in several studies since the middle 1950s, and in the late 1950s a new technique for studying bird migration was made available when a national network of approximately 50 weather surveillance radars (WSR-57) was established by the National Weather Service. By the mid-1960s these radars were being used to monitor the migration of birds at several locations on the northern coast of the Gulf of Mexico. With these powerful radars one could collect data 24 hours a day on the density, altitude, timing, direction, and geographical distribution of migration—thus eliminating much of the guesswork and shortcomings associated with earlier direct visual studies of migration. This radar network was updated in the early 1990s when new, sophisticated Doppler weather surveillance radars (NEXRAD, for *nex*t generation *rad*ar, or the WSR-88D) replaced the aging WSR-57 network. These new radars were not only better at detecting migrating birds in the atmosphere, but they displayed the relative density of the flights as well as the direction and speed of the migrants. In conjunction with the technological advances, the dramatic increase in the number of competent bird-watchers on the Gulf Coast produced considerable information on the distribution and timing of migration throughout the region. The information gathered from these sources provides the basis for the following overview of bird migration in relation to the Gulf of Mexico.

The Evolution of Trans-Gulf Migration

Why would small songbirds fly across the Gulf of Mexico on spring migration if they could go around it? This question has been posed many times and the answer relates to how the trans-Gulf flight strategy possibly evolved. Natural selection optimizes (maximizes) benefit of a strategy relative to the cost of that strategy. If two flight strategies (circum-Gulf versus trans-Gulf) have equal cost/benefit ratios, then either strategy will do about the same, and one would expect over time to see each one used about equally. But if the cost/benefit ratios differ between the flight strategies, then the strategy that produces the greatest benefit-to-cost ratio will be favored, and this strategy will predominate in subsequent generations of migrants. When this evolutionary argument is applied to the two migration strategies in question, we find that one (trans-Gulf migration) may sometimes be favored over the other (circum-Gulf). This conclusion is based on considerations of the energetics of the migrants making the flights, the amount of time required to complete the flights, and the importance of predation mortality along the migration routes.

Bird flight is an energetically costly endeavor, and in preparation for migration birds become *hyperphagic* (eat more) (see Chapter 2). The increased input of food results in fat accumulations that are deposited below the skin and in major muscle masses. During flight the fat deposits are used as fuel. Because birds crossing water barriers may encounter changes in weather (e.g., soaking rain, opposing winds), having a surplus of fuel is beneficial and very adaptive. It is now known that many

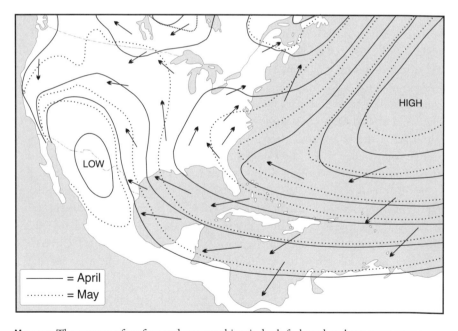

Map 3.2 The pattern of surface and geostrophic winds aloft, based on long-term average weather patterns for the months of April and May. The arrows show the surface wind directions and the geostrophic winds at approximately 2,500 feet above the ground flow, which follows isobars (lines of equal barometric pressure).

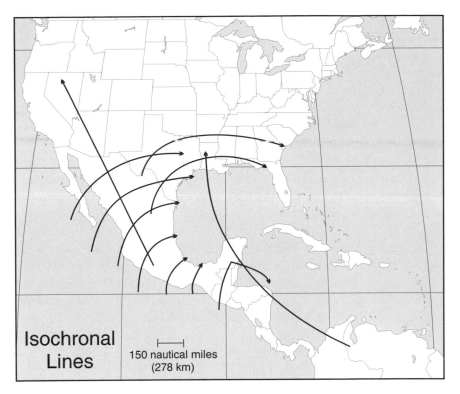

Map 3.3 The number of flight segments required to migrate from central Yucatán to southern Louisiana following a circum-Gulf strategy versus a trans-Gulf strategy.

long-distance, nonstop migratory flights cannot be completed even with a full fuel load unless the flight is made with assisting or following winds. One can think of these winds in a sense as fat or fuel equivalents. The more favorable the winds, the less fuel it takes to make the trip. It has been suggested and recently demonstrated that birds arriving on the breeding grounds with a surplus of fuel do better in reproduction than birds that arrive lean or in poor energetic condition. Thus, the presence of following winds becomes of paramount importance for birds crossing the Gulf. In spring the winds aloft over the Gulf of Mexico are for the most part very conducive to trans-Gulf migration (Map 3.2) and enable most birds to arrive on the northern coast of the Gulf in excellent physiological condition with fuel to spare.

As birds migrate northward overland in spring, they often traverse unfamiliar landscapes and must put down and spend the daylight hours in ecologically diverse stopover areas where predation pressures can be a significant problem. When one calculates the number of flight segments a typical nocturnal migrant songbird would make to go around the Gulf of Mexico versus flying directly across the Gulf, it becomes readily apparent that there is a significant time advantage and reduced predation pressure for those high-flying (and therefore less vulnerable) migrants crossing directly over the Gulf (Map 3.3). With favorable or following winds migrants can leave the central portions of the Yucatán Peninsula at the beginning of the night, fly out over the Gulf, and arrive on the northern Gulf Coast in the middle-to-late afternoon of the next day—a journey of approximately 450 nautical miles (834 km) that takes about 15 hours with an average ground speed of 30 nautical miles per hour (knots), or 15.4 m/s. A significant portion of the "fuel" for this journey comes from the assistance of the following wind. In contrast, a bird starting from the same place on the Yucatán Peninsula and going around the Gulf would take from five to six days to reach the same point on the northern Gulf Coast, because overland migratory flights do not continue into the daylight hours. Growing amounts of data show that migrants which arrive early on their breeding grounds and in good energetic condition are able to maximize their reproductive success. Clearly, if minimizing the length of the migratory journey, avoiding predators, and saving fuel are important and beneficial factors that facilitate a migrant's early arrival in good condition, then it is not surprising that trans-Gulf migration is a regular occurrence in spring despite the occasional catastrophes that befall migrants when strong cold fronts with stormy weather move southward over the Gulf.

The Characteristics of Spring Trans-Gulf Migration

The pioneering studies of trans-Gulf migration by George Lowery and Robert Newman at Louisiana State University (using moon-watching, vertical telescope, and censuses of birds at stopover sites) in the late 1940s and through the 1950s stimulated additional studies of migration across and around the Gulf in subsequent decades. In the 1960s, the use of weather surveillance radars along the northern Gulf Coast permitted the detection, monitoring, and quantification of arriving trans-Gulf flights in spring [10]. In addition, the establishment of banding sites and studies of migrant bird–habitat associations at isolated stopover sites greatly expanded our knowledge base of bird migration on the northern Gulf Coast [11]. All of these studies have led to the current understanding of the nature of trans-Gulf migration.

Seasonal Timing and the Patterns of Winds Aloft

Spring trans-Gulf migration begins typically as small flights in the first and second week of March, reaches a peak with very large flights in late April and early May, and

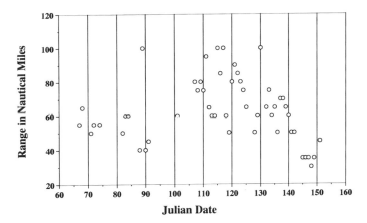

Figure 3.2 Seasonal pattern of the arrival of trans-Gulf migrations on the northern Gulf Coast. These data were gathered using the WSR-57 weather surveillance radar at Galveston, Texas, in the spring of 1990. The density of each trans-Gulf flight is measured in terms of the mean number of migrant flocks aloft in a 20° sweep of the 2° radar beam elevated 2.5°.

is essentially finished by the third week in May. Only rarely do very small flights continue until the end of May [12]. The largest trans-Gulf flights generally occur after 10 April (Julian date 100), and the flights begin to decline in size after 10 May (Julian date 130) (Fig. 3.2). The patterns of winds aloft over the Gulf of Mexico are critically important to the seasonal timing of trans-Gulf migration. In March, the winds over the Gulf are often influenced by continental polar air masses (anticyclonic systems over the southeastern United States), and winds blow from the east near the surface and aloft. Only when a return flow of maritime tropical air from the south occurs are conditions good for a south-to-north trans-Gulf crossing, and this event occurs aloft before it occurs on the surface. Consequently, winds aloft are generally more favorable for a Gulf crossing than are surface winds. As spring progresses, the number of days with good return flow increases. In April and May, the pattern of winds aloft becomes increasingly more favorable as cold fronts decrease in frequency [13].

The Daily Timing of Trans-Gulf Flights

Like the seasonal timing, the daily timing of trans-Gulf migration is dependent in large measure on the weather conditions associated with the journey. We know from the moon-watching data collected by Lowery and his colleagues that birds take off from areas south of the Gulf of Mexico just about the time it is dark and fly northward out over the Gulf. Although our information on weather conditions at the time of departures is limited, they likely occur in favorable weather (with assisting winds and no rain). Birds continue to cross the coastline and fly out over the Gulf for a period of almost eight hours, with peak numbers crossing the northern coast of Yucatán around midnight. The last birds in the movement could have come from as far as 280 km, or 150 nautical miles, to the south (e.g., Belize).

The arrival of trans-Gulf migration on the northern Gulf Coast typically occurs during the daylight hours, and the subsequent exodus of migrants from coastal stopover areas takes place in an explosive fashion within the first two hours of the evening (Fig. 3.3). The arrival time of a trans-Gulf flight on the northern Gulf Coast is strongly correlated with the speed of the southerly winds aloft and is also influenced by disturbed weather over the Gulf [12]. Typically, with 10–15 mph southerly winds (9–15 knots, or 5–8 m/s) the movements begin to arrive about 10:00 a.m. local time, reach peak densities in the midafternoon, and are largely finished by nightfall. The display of the migration on radar is easily recognized and can be discriminated from precipitation echoes (Plate 3.2). When the winds aloft are strong (30 mph, or 13 m/s), trans-Gulf migrations may begin to arrive on the northern Gulf Coast before dawn, peak before noon, and terminate by midafternoon. When migrants crossing the Gulf are displaced westward or eastward by side winds, the journey is lengthened and the arrival on the coast is delayed.

In favorable flying conditions, most of the trans-Gulf migrants generally fly over the coast and make landfall in forests inland from the coast (Fig. 3.4). After their

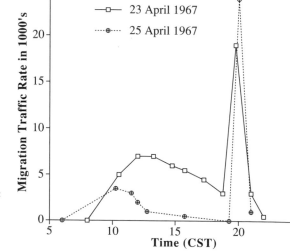

Figure 3.3 Daily temporal patterns of the arrival of trans-Gulf migration and the subsequent exodux from coastal stopover areas. The migration traffic is the number of birds crossing 1.6 km (1 mi) of front per hour.

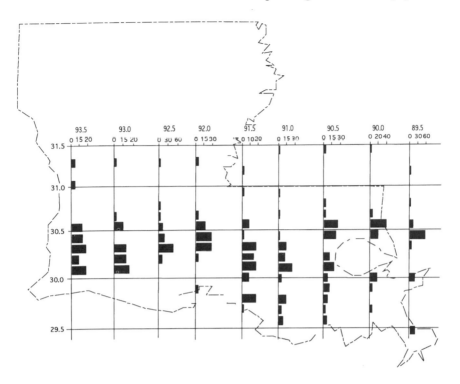

Figure 3.4 Landing zones of springtime migrants in southern Louisiana after trans-Gulf migration. The data are derived from 95 trans-Gulf flights recorded with the WSR-57 weather radars in New Orleans and Lake Charles. Latitudes and longitudes are graphed over a map of Louisiana. The bars represent percentages of the 95 flights that landed at that latitude.

flight across the Gulf of Mexico, one would expect that the migrants would rest for a few days before resuming their northward migration. This is typically not the case. Most of the migrants that land during the daylight hours depart the stopover areas 30 to 40 minutes after sunset on the same day (during nautical twilight) and begin another leg of their migratory journey. On the radar display the exodus is often spectacular as the density of migrants aloft increases rapidly (Plate 3.3). The duration of the exodus rarely exceeds three hours for radar stations on the central, northern Gulf Coast, but for stations on the Texas coast the movement of migrants toward the northeast continues for most of the night. These movements contain migrants that arrived from across the Gulf as well as migrants moving up from eastern Mexico [14]. The seasonal pattern of the exodus of trans-Gulf migrants from stopover areas on the northern Gulf Coast reflects the input pattern of trans-Gulf

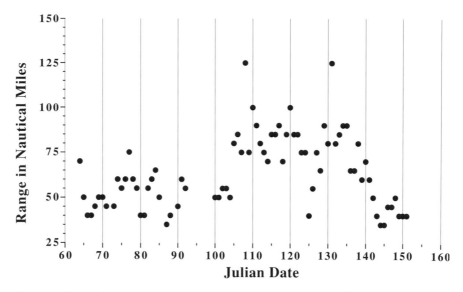

Figure 3.5 Seasonal pattern of the exodus of trans-Gulf migrants from "coastal" stopover areas on the northeastern Gulf Coast of Texas. These data were gathered using the WSR-57 at Galveston, Texas, in the spring of 1990. Note that the seasonal timing of the largest exoduses corresponds to that for the largest trans-Gulf migrations (*see* Fig. 3.2). The magnitude of a movement is proportional to the maximum range to which its echo pattern extends on the radar screen.

migrations (Fig. 3.5). In March, the magnitude of the flights is small usually, but after 10 April (Julian date 100) the flights increase in magnitude. The size of the exodus begins to decline after 10 May (Julian date 130), and by the end of May the pulse of an exodus is barely visible on radar. When a trans-Gulf flight does not arrive, the size of the exodus that evening drops considerably; and when a trans-Gulf flight arrives after dark, the "exodus" may be prolonged for several hours.

The Effects of Unfavorable Weather Conditions

When conditions over the Gulf are less favorable (rain and unfavorable winds) the pace of migration is slowed and the arrival of trans-Gulf migrants on the Gulf Coast may be delayed until after dark. Only during adverse weather (hard rain or strong north winds) do substantial numbers put down in coastal woodlands (Fig. 3.6). On rare occasions powerful cold fronts move southward over the Gulf in late April and early May with frontal storms and subsequent north winds. Trans-Gulf migrants in-

tercepted by these fronts must fly northward against the wind, often making little
headway. They are forced to fly for longer than usual and in some cases use all of
their on-board fuel. Exhausted and out of fuel they put down on the first available
"land." Under these circumstances, spectacular fallouts of migrants take place on
offshore oil rigs and fishing boats in the northern Gulf, and in coastal woodlands
along the Gulf Coast (Plate 3.4). Many of the birds recovered from the oil rigs and
fishing boats have concave breast muscles and the keel of their sternum is like a knife
blade, indicating that they have catabolized breast muscle for fuel. Many of these
migrants, like the migrants that fall into the waters of the Gulf, do not survive.
Those that make it to the coastal woodlands must replenish their fuel supply and
delay further migration until their energetic condition improves. When cold fronts
are weak and shallow, most trans-Gulf migrants continue flying north in the
southerly airflow above the frontal boundary. Thus in spring it is possible to have a
weak cold front with no fallout of migrants: in such cases, the trans-Gulf migrants
have overflown the front to reach their inland stopover destinations. Most of the

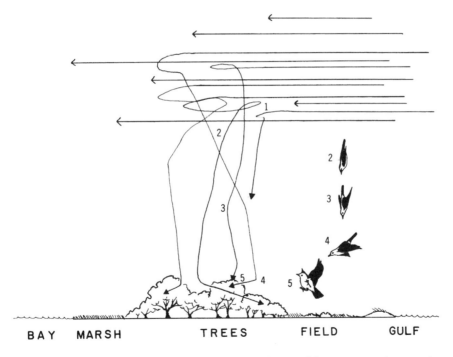

BAY   MARSH   TREES   FIELD   GULF

Figure 3.6 The flight behavior of trans-Gulf migrants during a fallout at a coastal site on the
northern coast of the Gulf of Mexico. Although many birds continue farther inland to more
extensive forest, some birds put down in the first available woodland.

birds that put down on offshore oil rigs and in coastal woodlands *when flying conditions are good* are likely to be physiologically stressed, suffering from dehydration, or very low on fuel [15, 16].

The Altitude of Trans-Gulf Migration

The altitude of trans-Gulf migration as it arrives on the northern coast of the Gulf is considerably higher than that of most migratory movements of songbirds over land (Fig. 3.7). On many occasions one cannot look overhead in a clear sky and see the arriving migrants even with a telescope: the birds are just too high to see against a bright blue sky. On other occasions the birds are arriving above a solid cloud layer. Were it not for radar, these movements would likely go undetected. Trans-Gulf migrants tend to stay above the cumulus clouds that begin to form just north of the coastline; the migrants build higher in altitude as the clouds move inland on southerly winds. In many instances, the thickness of the layer of trans-Gulf migrants aloft is on the order of 5,000 to 6,000 feet (1,500–2,000 m), and the altitudinal distribution of flocks in a flight may range from 1,900 to 8,200 feet (580–2,500 m). The altitudinal distribution is strongly dependent on the patterns of the winds aloft. If the most favorable southerly flows are at low altitudes, then the distribution of trans-Gulf migrants is skewed to the lower altitudes. If the most favorable southerly winds

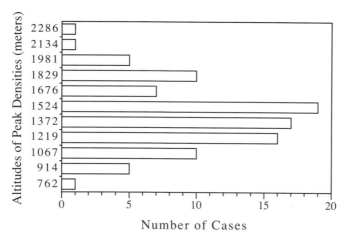

Figure 3.7 The altitudinal distribution of peak densities of arriving trans-Gulf migrations at New Orleans and Lake Charles, Louisiana (both stations combined). The altitudes were measured with wsr-57 radars from 1965 through 1967.

are at higher altitudes, then the migrants will fly at those altitudes. In one instance, trans-Gulf migrants were flying at altitudes of 12,000 to 15,000 feet (3,657–4,572 m) as they moved inland above a shallow cold front that had penetrated the northern Gulf. When powerful cold fronts move well into the Gulf and trans-Gulf migrants cannot fly over the altitudinal zone of adverse winds, the birds lower their flight altitude and fly just above the waters of the Gulf. On these occasions they may actually be flying below the coverage of the radar, but they can be detected readily when they depart the "coastal" areas during their exodus. Once over land and suitable habitat, trans-Gulf migrants do not fly against strong northerly winds.

The Geographical Extent of Trans-Gulf Migration

Early research on trans-Gulf migration may have produced the misconception that most of the birds that cross the Gulf of Mexico in spring come from the Yucatán Peninsula, for the only direct visual information on migrants flying out over the

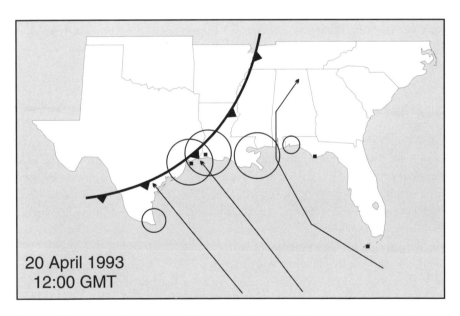

20 April 1993
12:00 GMT

Map 3.4 The geographical pattern of the arrival of a trans-Gulf migration on the northern Gulf Coast. The arrows indicate the direction of the winds aloft; the circles indicate the maximum range to which the echo pattern from trans-Gulf migrants extended on the radar screen. The magnitude of the migratory movement is proportional to the radius of the circle.

Figure 3.8 The black-whiskered vireo occurs in Louisiana when drifted by winds westward from its usual path of migration. (Photo by L. Page Brown for Cornell Laboratory of Ornithology)

Gulf at night came from Yucatán. The current view, greatly improved by the use of weather surveillance radar, is that migrants cross the Gulf from source areas in the Caribbean, Yucatán, and portions of Mexico to the south and west of the Bay of Campeche. The geographical patterns of the source areas are reflected in the geographical patterns of the arrivals on the northern Gulf Coast in relation to the synoptic weather patterns over the Gulf. Typically the "main bang" of the trans-Gulf flight covers a 500-mile segment of coast, with smaller amounts of migration occurring on either side (Map 3.4). Year in and year out, the two coastal stations that detect the greatest amount of trans-Gulf migration are Houston, Texas, and Lake Charles, Louisiana. Prevailing weather conditions over the Gulf generally vector the greatest concentrations of migrants aloft over the Gulf to these areas. When the major wind patterns over the Gulf shift, the area that receives the greatest migration shifts accordingly. If strong winds from the southwest flow across the Gulf, the migrants will be shifted eastward and dense migrations will arrive on the northeastern Gulf Coast. Such conditions often bring a few "western" migrants to these areas (e.g., western kingbird, mourning warbler). Likewise, when wind conditions over the Gulf are from the east and southeast, most of the trans-Gulf migrants are displaced westward and make landfall on the upper Mexican and lower Texas coasts. On these occasions, eastern migrants that typically migrate up the Florida Peninsula may be found on the upper Texas coast (e.g., Cape May warbler, black-throated blue warbler). These birds drifted westward, missed the Florida Peninsula, and moved out over the Gulf. For many years, the earliest spring record of the black-whiskered vireo (Fig 3.8) in the United States was from Grand Isle, Louisiana, where it is very rare, instead of from Florida, where it breeds—clearly a case of westward drift!

The Pattern of Fall Trans-Gulf Migration

Although I have emphasized spring migration largely because more work has been done at that season following the Lowery and Williams debates, the issue of trans-Gulf migration in the fall is equally important. Interestingly, there has been little debate about fall migration. It has been generally accepted that migrants stimulated by cold fronts in fall leave stopover areas on the northern coast, move out over the Gulf with following winds, and complete a crossing. The major problem with this idea is that powerful cold fronts that are able to penetrate the Maritime tropical air over the Gulf are decidedly rare during most of the fall migration (early August through mid-November), and the prevailing weather patterns over the Gulf in fall (south and easterly winds) are not particularly favorable for a direct north–south crossing.

In a continentwide study involving moon-watching by hundreds of cooperators, Lowery and Newman showed the relationship between favorable winds for trans-Gulf migration and the departure of migrants from the northern Gulf Coast on a trans-Gulf crossing for four nights in October (between twilight on 1 October and dawn on 5 October in 1952) [17]. The largest trans-Gulf movements during the study were stimulated by a cold front that moved out over the Gulf on 2 October.

A similar pattern of migration on the northern coast of the Gulf in fall was reported by Ken Able when he studied nocturnal songbird migration in southwestern Louisiana for an entire fall season using radar and a vertical ceilometer beam [18]. He found that the migrants generally flew with the wind, regardless of the direction, and this meant that trans-Gulf flights from that area were relatively infrequent. Reversed migrations (northward) in the fall were common. However, the prevailing winds were from the northeast at this location and resulted in a strong net flow of migrants in a southwesterly direction that would take them along the Texas coast on a circum-Gulf route rather than across the Gulf.

On the other hand, Bill Buskirk looked for the fall arrival of trans-Gulf migrants on the northern coast of Yucatán and found small incoming flights almost daily. Although these flights could not have come from areas on the Louisiana and Texas coasts, they could have originated from the Florida coast, where in the fall winds from the north and east are also regular and could carry the birds across the Gulf to Yucatán [19].

As it now stands, we know considerably more about spring migration than about fall migration in the region of the Gulf of Mexico. The importance of conducting additional migration studies along the northern Gulf Coast and over the northern Gulf in fall is obvious. Preliminary findings using a network of wsr-88d radars on the northern Gulf Coast during the fall seasons of 1995 and 1996 reveal unexpected patterns of trans-Gulf migration in the fall. In the absence of cold fronts, the migrants appear to move southward following the Texas coast and also down the peninsula of Florida. Many of the flights from the northern Gulf Coast, following cold fronts, appear to be directed toward the southwest such that the migrants reach the south Texas coast and the upper Mexican coast. On such occasions, migrants arrive on the south Texas coast *from the northeast, crossing the Gulf during daylight hours* — a pattern not all that different from what happens with trans-Gulf migration in the spring!

Trans-Gulf Flights and Migrant Populations

Over the last decade, there has been considerable debate about the population trends of Neotropical migrants — birds that breed in the United States and Canada and winter south of the border. The overwhelming majority of neotropical migrants that breed in the grasslands and woodlands of central and eastern North America must pass through the northern Gulf Coast on their northward journey from the tropics. The conventional wisdom among veteran bird observers along the northern Gulf Coast is that both the frequency and the magnitude of spring fallouts have declined markedly over the past three decades. Because fallouts are dependent on so many variables, they may be an unreliable estimate of what is really going on in populations of trans-Gulf migrants. In an effort to quantify the long-term trends of Neotropical migrant populations, analyses of the North American Breeding Bird Survey database (1966–present) were undertaken in the late 1980s. The results suggested that populations of migrants breeding in the eastern deciduous forest generally increased from 1966 through the 1970s and declined during the 1980s [20]. However, subsequent analyses by different investigators using the same data suggested a complex pattern of regional differences in population trends, with some species increasing and others decreasing (see Epilogue). There was an obvious need for an analysis based on a different type of data.

Because routine filming and archiving of weather radar displays has been carried out since the wsr-57 unit came on line in 1957, an enormous database exists that potentially can document trends in the occurrence of trans-Gulf migration for a period of more than 3 decades. In an effort to see if the declines might be detectable in the archived films from the network of weather radars on the northern Gulf Coast, I conducted a preliminary analysis of films from Lake Charles for the spring seasons of 1965 through 1967 and 1987 through 1989. The results supported the notion that populations of Neotropical migrants had declined. The percentage of spring days with arriving trans-Gulf flights in the latter years was only about 50% of that in the earlier ones. Flights earlier in the migration season seemed to be especially reduced, suggesting that perhaps early migrants (e.g., species that breed primarily in the southeastern states) (Plate 3.5) have been impacted more severely than May migrants that breed for the most part farther north [21]. It is important to emphasize that this analysis was preliminary and based on only 1 radar station. Recent analyses of the archived radar films from Galveston, Texas, for the same period suggest that the pattern observed at Lake Charles may not be representative of other radar stations along the northern Gulf Coast, but the quality of the Galveston film archive is poor and plagued by missing data and inappropriate radar settings. Additional analyses of the WSR-57 archived films are under way. In the meantime, the new network of state-of-the-art Doppler weather radars is in place along the coast, and the data from these radars are now being archived for use in monitoring the long-term patterns of trans-Gulf migration. These new radars are more powerful and sensitive than the retired WSR-57 units, and the quality of the archive will allow quantitative monitoring of the health of the trans-Gulf migration system well into the future.

References

1. Cooke, W.W. 1904. Some new facts about the migration of birds. *In* Yearbook of Department of Agriculture for 1903. Washington, DC: GPO. p. 371–386.
   One of numerous early accounts by Cooke. Contains his assertion that birds migrate across the Gulf of Mexico.
2. Williams, G.G. 1945. Do birds cross the Gulf of Mexico in spring? Auk 62:98–11.
3. Lowery, G.H. Jr. 1945. Trans-Gulf spring migration of birds and the coastal hiatus. Wilson Bull. 57:92–121.
4. Lowery, G.H. Jr. 1946. Evidence of trans-Gulf migration. Auk 63:175–211.
5. Williams, G.G. 1947. Lowery on trans-Gulf migration. Auk 64:217–238.
6. Williams, G.G. 1950. The nature and causes of the coastal hiatus. Wilson Bull. 62:175–182.
   References 2 through 6 are the exchanges in the literature that constituted the circum-Gulf versus trans-Gulf migration controversy.
7. Lowery, G.H. Jr. 1951. A Quantitative Study of the Nocturnal migration of Birds. Univ. Kansas Publ., Mus. Nat. Hist. Volume 3.
   The original description of the moon-watching technique and its application to the study of migration, especially in the Gulf region.
8. Lowery, G.H. Jr., and R.J. Newman. 1954. The birds of the Gulf of Mexico. *In* Gulf of

Mexico, Its Origin, Waters and Marine Life. Fisheries Bull. 89. Washington, DC: U.S. Fish & Wildlife Serv. Volume 55.

Results of moon-watching and daytime telescopic watches.

9. Stevenson, H.M. 1957. The relative magnitude of the trans-Gulf and circum-Gulf spring migrations. Wilson Bull. 69:39–77.

Analysis of trans-Gulf versus circum-Gulf migration patterns based on the timing and relative abundance of species' migrations in the Gulf region.

10. Gauthreaux, S.A. Jr. 1970. Weather radar quantification of bird migration. BioScience 20:17–20.

The original description of the method for estimating quantities of migration from the displays on WSR-57 weather radar.

11. Moore, F.R., P. Kerlinger, and T.R. Simons. 1990. Stopover on a Gulf Coast barrier island by spring trans-Gulf migrants. Wilson Bull. 102:487–500.

Behavior, abundance, habitat selection, and energetics of migrants stopping on Mississippi coastal islands.

12. Gauthreaux, S.A. Jr. 1971. A radar and direct visual study of passerine spring migration in southern Louisiana. Auk 88:343–365.

The classic radar study of spring trans-Gulf migration; discusses daily and seasonal timing, flight directions, weather effects, and more.

13. Duncan, R.A. 1994. Bird Migration Weather and Fallout: Including the Migrant Traps of Alabama and Northwest Florida. Published by R.A. Duncan, 614 Fairpoint Dr., Gulf Breeze, FL 32561.

A practical guidebook for birders who wish to observe spring migrant fallouts on the northern Gulf Coast; detailed information on when and where to look for migrants.

14. Forsyth, B.J., and D. James. 1971. Springtime movements of transient nocturnally migrating landbirds in the Gulf coastal bend region of Texas. Condor 73:1

The relative abundance and seasonal timing of migrant species on the Texas coast.

15. Spengler, T.J., P.L. Leberg, and W.C. Barrow Jr. 1995. Comparison of condition indices in migratory passerines at a stopover site in coastal Louisiana. Condor 97:438–444.

An evaluation of several methods for estimating the fat stores of migrants.

16. Leberg, P.L., T.J. Spengler, and W.C. Barrow Jr. 1996. Lipid and water depletion in migrating passerines following passage over the Gulf of Mexico. Oecologia 106:1–7.

A study of fat depletion and dehydration in migrants arriving in Louisiana after a trans-Gulf flight.

17. Lowery, G.H. Jr., and R.J. Newman. 1966. A continentwide view of bird migration on four nights in October. Auk 83:547–586.

A simultaneous view of migration based on a continentwide moon-watching network.

18. Able, K.P. 1972. Fall migration in coastal Louisiana and the evolution of migration patterns in the Gulf region. Wilson Bull. 84:231–242.

A radar and visual study of fall migration on the northern Gulf Coast as related to weather, especially wind patterns.

19. Buskirk, W.H. 1980. Influence of meteorological patterns and trans-Gulf migration on the calendars of latitudinal migrants. *In* Migrant Birds in the Neotropics: Ecology, Behavior,

Distribution, and Conservation (A. Keast and E. Morton, eds.). Washington, DC: Smithsonian Inst. Pr.

Examines the seasonal timing of spring and fall trans-Gulf migration in relation to weather and wind patterns.

20. Robbins, C.S., J.R. Sauer, R.S. Greenberg, and S. Droege. 1989. Population declines in North American birds that migrate to the Neotropics. Proc. Nat. Acad. Sci. U.S.A. 86:7658–7662.

The classic paper which first called attention to population declines based on data from the North American Breeding Bird Survey.

21. Gauthreaux, S.A. Jr. 1992. The use of weather radar to monitor long-term patterns of trans-Gulf migration in spring. *In* Ecology and Conservation of Neotropical Migrant Landbirds (J.M. Hagan III and D.W. Johnston, eds.). Washington, DC: Smithsonian Inst. Pr.

Compares the frequency of trans-Gulf spring arrivals in the late 1960s with the frequency in the 1980s; based on archival films from the WSR-57 radar.

Plate 1.1 The pine grosbeak is typical of irruptive species, emigrating from its boreal forest home when forced to do so by failure of its food supply. (Photo by K.P. Able)

Plate 1.2 A pair of blackcaps (male on left, female on right) feed their brood. (Photo by G. Moosrainer, courtesy of P. Berthold)

Plate 1.3 The house finch, introduced to Long Island, New York, around 1940, has become migratory during the subsequent 50 years or so. (Photo by T. Vezo/Vireo)

Plate 1.4 Cliff swallows are renowned for the precise timing of their spring return to the mission at San Juan Capistrano. In fact, many migrants are just as timely in their migratory schedules. (Photo by K.P. Able)

Plate 2.1 Strong, fast-flying shorebirds like the whimbrel are capable of making long nonstop flights. (Photo by K.P. Able)

Plate 2.2 Geese (such as these snow geese), cranes, and cormorants often fly in formations that result in energy savings for all individuals except the leader. The leader changes periodically. (Photo by K.P. Able)

Plate 2.3 This wood thrush is equipped with a small radio transmitter so that it may be followed while returning to its nest. The transmitter weighs about 2 g and is glued to the bird's back. The wire extending backward toward the tail is the transmitter antenna. (Photo by K.P. Able)

Plate 2.4 Birds are placed in small cages within these large coils, which change the earth's magnetic field. Copper wire wrapped on the wooden framework carries a small amount of current. The amount of current and the orientation of the coils determines the characteristics of the magnetic field created within the coils. (Photo by K.P. Able)

Plate 3.1 The male painted bunting is one of the most colorful trans-Gulf migrants. Sometimes seen in mixed flocks with indigo buntings, it is a stunning sight. (Photo by B. Schorre/Vireo)

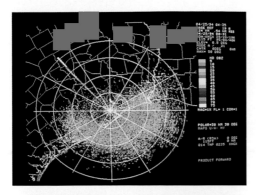

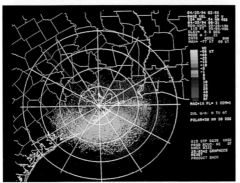

Plate 3.2 The arrival of flocks of trans-Gulf migrants around Houston, Texas, on 25 April 1994 at 00:31 GMT as detected by the WSR-88D weather surveillance radar. *Top:* base reflectivity image showing the density of migrants in the atmosphere. The color DbZ scale indicates the "density" of the birds in the atmosphere. *Bottom:* base velocity image showing the direction and radial velocities of movement. Targets colored green have radial velocities moving toward the radar while those colored red have radial velocities moving away from the radar. Note where the birds are landing (where targets disappear). (Photo by S. A. Gauthreaux Jr.)

Plate 3.3 An exodus of trans-Gulf migrants from stopover areas around Houston, Texas, on 25 April 1994 at 01:59 GMT as detected by the WSR-88D weather surveillance radar. *See* the legend to Plate 3.2 for explanation. (Photo by S. A. Gauthreaux Jr.)

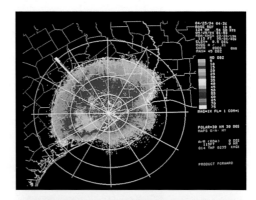

Plate 3.4 Scarlet tanagers spend the winter in South America. In spring, their trans-Gulf migration may actually be a nonstop continuation of a flight across the Caribbean. (Photo by W.A. Paff for Cornell Laboratory of Ornithology)

Plate 3.5 The hooded warbler (female shown here) breeds primarily in the south-eastern United States and arrives with the earlier contingent of trans-Gulf migrants in the spring. (Photo by B. Schorre/Vireo)

Plate 4.1 Though small and scraggly, the trees that make up the cheniers of southwestern Louisiana are the first that a trans-Gulf migrant sees when it finally reaches the coast. (Photo by K.P. Able)

Plate 4.2 Weighing, measuring, scoring the fat levels of, and banding migrants are integral parts of research on the stopover ecology of migrants. Here Paul Kerlinger processes a red-eyed vireo. (Photo by F.R. Moore)

Plate 4.3 This red-eyed vireo has been fitted with colored plastic leg bands so that it can be identified individually with binoculars. Information on the movements and behavior of this marked bird can be obtained without having to capture it a second time. (Photo by F.R. Moore)

Plate 4.4 The gray-cheeked thrush is a very long-distance migrant. It is possible that these thrushes may cross the Gulf of Mexico and the Caribbean in a single flight. They are one of several species that we have studied intensively in the Louisiana cheniers. (Photo by G. Bailey/Vireo)

Plate 4.5 Migrating raptors like this merlin pose a threat to stopover migrants in the cheniers. (Photo by J. Heidecker for Cornell Laboratory of Ornithology)

Plate 4.6 This white-eyed vireo became a part of our study when it flew into a mist net. (Photo by F.R. Moore)

4

FRANK R. MOORE

Neotropical Migrants and the Gulf of Mexico: The Cheniers of Louisiana and Stopover Ecology

In the late 1960s, a group of us scouting for a big day in Cameron Parish, Louisiana, were looking over some ducks resting offshore in the Gulf. Suddenly there appeared, flying directly toward us (and the shore), a bizarre-looking creature. Flapping madly, only inches above the water, barely making it over the waves, legs dangling—nothing clicked. So out of context was it, that only when it got close enough that the legs could be seen to be yellow did it dawn on us that here was a purple gallinule arriving after a trans-Gulf flight. Arriving out of gas. It crossed the beach below eye level and when it reached the low, grass-covered dunes it crash-landed, did an inelegant somersault, and ran off into the brush. Just beyond the dunes lay endless marshland, perfect gallinule habitat. This migrant made it, but barely.

Songbirds reaching the coast after eighteen hours or so of nonstop flight face a less hospitable environment. Most of the immediate Coastal Plain is wet prairie and marsh. Trees occur in small, isolated patches on the almost imperceptibly higher ground of ancient beachlines. Even these are mostly poor, scraggly excuses for trees. No wonder that if the flying conditions have been favorable, most arriving migrants bypass the coastal woodlots and cheniers to continue inland to greener pastures—the coastal hiatus that Sid Gauthreaux described in Chapter 3. But some birds land at the coast even on good flight days. Presumably these are individuals low on fuel or water, or in some other way stressed physiologically. And when the weather is bad or the birds have had to fight head winds to make landfall, the isolated coastal woodlands can fill up with vast numbers of migrants—the classic Gulf Coast fallout. These areas provide ideal sites in which to study the behavior and ecology of birds trying to cope with a major life problem: how to rest and refuel as quickly and efficiently as possible so that migration can continue.

Since locating at the University of Southern Mississippi more than twenty years ago, Frank Moore has taken advantage of the opportunity to ask how migrants on stopover deal with these challenges. He and his students have studied migrants on dry offshore islands and wet oak cheniers, both in spring after a trans-Gulf migratory flight and in the fall as some of the birds prepare for a long migration. Most birders who are lucky enough to encounter a fallout in coastal Louisiana will have the time and inclination only to identify the species they see and perhaps to estimate their numbers. But there is a life-and-death drama playing out here. Follow Frank into the chenier and find out what the birds are doing. —K.P.A.

Peveto Beach, East Jetty Woods, and Johnsons Bayou are synonymous with the spring arrival of songbirds along the northern coast of the Gulf of Mexico. Each is a coastal woodland, or "chenier," in Cameron Parish, Louisiana, where weary migrants find temporary respite after a potentially hazardous trans-Gulf flight, where bird-watchers can sometimes see hundreds of warblers, vireos, tanagers, and thrushes of several dozen species on an afternoon walk through the woods, and where my graduate students and I have studied the stopover biology of these long-distance travelers since 1985 (Figure 4.1).

The French word *chenier* means "place of oak" and refers to forested habitat on relic beach ridges that run roughly parallel through the marsh and prairie of the Chenier Plain, which stretches from East Texas through much of coastal Louisiana. Cheniers in Louisiana are typically dominated by live oak and hackberry, but include toothache-tree, red mulberry, and honey locust. The shrub understory is a mixture

Figure 4.1 Peveto Beach Woods, Cameron Parish, Louisiana, from the air. When the Baton Rouge Audubon Society preserved this unique woodland habitat, they established the first chenier sanctuaries for migratory birds: the Holleyman-Sheely Sanctuary (1984) and the Henshaw Sanctuary (1989). (Photo by F. R. Moore)

of hackberry saplings, yaupon, palmetto, sweet acacia, and numerous vines such as honeysuckle, poison ivy, greenbrier, and grape vine.

Although their vegetation is not spectacular or especially lush, these coastal woodlands provide important stopover habitat for Nearctic-Neotropical landbird migrants [1] (Plate 4.1). They represent a last possible stopover before fall migrants make a nonstop flight of 18 to 24 hours and greater than 600 miles (965 km), and they are the first possible landfall for birds returning north in spring. In spring alone, more than threescore species regularly stop over in coastal cheniers, sometimes in vast numbers, following trans-Gulf migration. Although the number of species varies daily, their diversity increases in mid-April and averages about 25 species per day through the first part of May. The number of individuals also varies from day to day, though peak numbers are evident during late April through the 2nd week in May. Males typically appear earlier than females for most species, and older males often precede younger, second-year males during spring passage.

Some of the birds that stop to rest and to forage in coastal cheniers will avoid predators, find food, replenish fat stores, and get on with their northward journey;

others will be less successful. Visualize a red-eyed vireo gleaning small caterpillars from the edge of hackberry leaves in the middle of the long, narrow chenier near Johnson's Bayou, Louisiana. Now consider the many "decisions" it must make in response to the problems encountered en route. Besides the energetic cost of transport, it must adjust to unfamiliar surroundings, must balance conflicting demands between predator avoidance and food acquisition, must compete with other migrants and resident birds for limited resources, must cope with unfavorable weather, and must correct for orientation errors [2, 3, 4]. How well it solves those problems will determine the success of its migration: a successful migration is ultimately measured in terms of survival and reproductive success.

What Habitat to Choose?

One of the first "stopover" decisions a migrant must make revolves around where to settle upon reaching the northern coast of the Gulf of Mexico. Recognition of a high-quality habitat where fat stores can be safely and rapidly replenished is critical to a successful migration. Yet, when a red-eyed vireo stops during passage, it almost invariably finds itself in unfamiliar surroundings at the very time energy demand is high. How does it assess habitat quality? What cues are used when deciding to settle in a particular place? Perhaps it is genetically programmed to respond to simple structural features of the vegetation. The presence of other migrants may hint at habitat quality, or maybe it tries to gauge food abundance and predation pressure by sampling habitat directly. Whatever the cues, time is of the essence.

Upon arrival, birds move quickly from treetop to treetop or from shrub to shrub within the chenier, often in loose mixed-species flocks, giving the impression that they are assessing habitat. Movement declines sharply within an hour of arrival and, once "settled" in chenier habitat, migrants often forage alone. If seen together, flocks are usually small and often composed of a single species. For example, Tennessee warblers, indigo buntings, and orchard orioles, which rarely forage alone during passage, usually associate with others of their own kind. Red-eyed vireos are often seen in groups of two, which is not uncommon among other species during stopover (Plate 4.2).

What to Eat?

Possibly the single most important constraint during migration is the need to acquire enough food to meet energetic requirements, especially for those long-distance migrants which must overcome a barrier like the Gulf of Mexico. Long-distance migrants like the red-eyed vireo develop ferocious appetites in order to deposit energy stores in the form of fat that can amount to 30% to 50% of their live body mass. They rely on those fat stores to fuel their long migratory flights [2, 5]. Therefore, it is not surprising to find many fat-depleted individuals among the birds that stop over in coastal woodlands after a trans-Gulf crossing. For example, more than a third of all thrushes captured at Peveto Beach Woods during one spring carried no visible subcutaneous fat, and the body mass of nearly a quarter of the thrushes was equal to or below their typical fat-free weight. Evidently, some birds had begun to break down muscle tissue as an emergency source of energy.

Although it is safe to assume that mass differences among birds reflect differences in fat stores, other factors contribute to the emaciated condition of some migrants. Long, nonstop flights over ecological barriers may dehydrate birds and thus depress body mass. Body mass also declines if muscle tissue is broken down to provide energy (*catabolized*) during long, nonstop flights.

The consequences of arriving in a fat-depleted condition are several, and none is favorable. First, lean migrants have a smaller "margin of safety" to buffer the effect of rainy, cold weather on the availability of food supplies during stopover. Second, efforts to satisfy energy demand may expose hungry migrants to increased risk of predation. Third, if lean birds remain longer at a stopover site than birds in better energetic condition and do not make up the lost time, they will arrive later on the breeding grounds. Birds that arrive late may experience trouble finding a good territory or attracting a mate.

Birds that rapidly restore fat loads to levels appropriate for the next leg of their journey will minimize the time spent en route, and field studies show that free-ranging migrants are capable of restoring fat loads at rates approaching 10% of their lean body mass per day [1]. When food is plentiful, even an inefficient forager may have few problems finding enough food to deposit fat stores rapidly. But when food is scarce or when faced with heightened energy demand in anticipation of a long, nonstop flight over the Gulf of Mexico, good foraging decisions may make the difference between life and death. Where to search for food and what items to include in the diet are important questions for a lean migrant that is pressed for time.

Moreover, plasticity inherent in the foraging behavior of migratory songbirds may allow a quick response to depleted fat stores. Field studies in Peveto Beach Woods (Plate 4.3) revealed that lean red-eyed vireos searched for food in a wider variety of habitats, sped up the rate at which they looked for food, and increased their foraging repertoire to include more than the usual gleaning maneuver in contrast to fatter birds. Lean individuals were frequently observed performing rather un-vireo-like behavior, such as hawking prey on the wing, and were sometimes seen hovering over foliage in search of food. As a consequence of these changes, lean birds were more likely to gain fat during stopover than birds carrying unmobilized stores [6]. If energetically stressed migrants increase their rate of energy acquisition, a favorable energy budget is achieved more quickly, length of stay decreases, and the speed of migration increases.

Although birds do attempt to replenish depleted fat stores during stopover, excessively large fat loads should not be expected following trans-Gulf migration even under the best of conditions. Because flight costs increase as the fat load accumulates, a migrant's flight range will be reduced as a function of its fat load [5]. Moreover, the large fat stores that provide a margin of safety during the long trans-Gulf flight would not be necessary for overland flight. Heavy fat loads may also impose additional costs during migration. Besides the energetic cost of accumulating extra fat, the added mass adversely affects flight performance and decreases maneuverability, possibly increasing the risk of predation.

Competition

As if it were not enough that a red-eyed vireo must find adequate food in an unfamiliar place, competition with other migrants and residents can become a serious matter. When migrants with similar food requirements and high energy demand are locally concentrated in an unfamiliar chenier, competition for available food can seriously reduce the rate at which each can satisfy its energy demand. Regardless of the absolute availability of food, if food is limited relative to energy demand, a reduced rate of food intake decreases the probability that a migrant will meet energetic requirements and successfully complete its migration.

The evidence for competition among migrating songbirds is largely circumstantial, although birds have been observed defending territory during stopover. During a study of en route competition, one of my graduate students and I prevented migrants that stopped in a Louisiana chenier from feeding on caterpillars that defoliate the hackberry trees. We wanted to see if the presence of potential competitors really depressed food abundance for other migrants, so we draped fine-mesh crop netting loosely around parts of several trees, which kept the birds away from the foliage (and their food) but permitted caterpillars to move freely about the tree. Numbers of caterpillars within our exclosures were higher than numbers in similar areas without the crop netting [7]. Thus, feeding by migrants depresses food resources for subsequent arrivals, setting the stage for competition en route. We were not surprised when we found that the rate at which several species replenished fat stores slowed considerably when the density of birds in the chenier was high (Fig. 4.2).

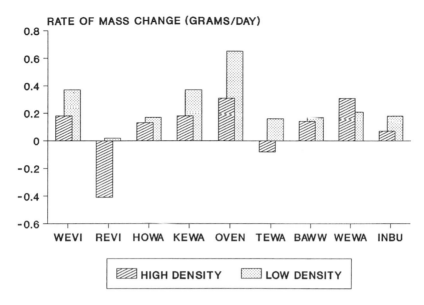

RATE OF MASS CHANGE (GRAMS/DAY)

Figure 4.2 Rate of body mass change for several species of Neotropical landbird migrants on days when more than 100 migrants stopped over at Peveto Beach Woods and when fewer than 50 migrants stopped. Species shown are the WEVI (white-eyed vireo), REVI (red-eyed vireo), HOWA (hooded warbler), KEWA (Kentucky warbler), OVEN (ovenbird), TEWA (Tennessee warbler), BAWW (black-and-white warbler), WEWA (worm-eating warbler), and INBU (indigo bunting).

Predation

Mortality associated with trans-Gulf migration and stopover in coastal habitat manifests itself in several ways: death during a trans-Gulf flight, starvation during the stopover following a trans-Gulf flight, and predation by raptors both in spring and fall. On daily walks along a stretch of beach during one spring field season, I came across nearly 100 dead birds of 20 different species—a conservative "sample" because carcasses are eaten by fish, crabs, gulls, and other scavengers. It may be interesting to note that two of the frequently found species, the gray-cheeked thrush (Plate 4.4) and the scarlet tanager, winter in South America and often arrive along the northern coast of the Gulf of Mexico in a fat-depleted condition. It makes one wonder if these birds are making nonstop flights from the northern coast of South America to the southern coasts of temperate North America, traversing both the Caribbean *and* the Gulf.

Coastlines often concentrate raptors during their migration [8], and several hawks that cross the Gulf of Mexico in both spring and fall occur frequently in cheniers and on barrier islands that parallel the northern Gulf of Mexico. It is not unusual for merlins and peregrine falcons to spend several consecutive days in a chenier, frequenting the same perch (Plate 4.5). Some bird-eating hawks probably migrate along coasts because of the seasonal concentration of potential prey, especially energetically stressed birds which make easy prey. It may not be coincidental that coastal concentrations of fall migrants are often highly skewed toward young, first-year birds, given the high number of bird-eating raptors seen in those areas: perhaps the experienced adults avoid coasts.

A real dilemma arises when the best feeding areas are also the most dangerous. A migrant must decide how to trade off energy gain against mortality risk, and it would not be surprising if hungrier (fat-depleted) birds were more likely to accept risk in order to meet energetic requirements. This trade-off may depend on age and experience. If dominant birds are older and fatter, they should be less hungry than young, leaner subordinates. Clearly, the problem of satisfying energy demand during migration is not simply a matter of finding enough food [3].

Water economy might constrain migratory range and could explain why some individuals stop despite sufficient fat stores for continued migration. Migrants might suffer muscular fatigue during a sustained flight over ecological barriers and might stop over to metabolize lactate and "repay" their oxygen debt, regardless of their fat status. Stopover may also be required for tissue repair if migrants are forced to catabolize too much muscle tissue or if muscle fibers are damaged during a long, sustained flight. In either case, a safe place to rest may be as important a determinant of site quality as food availability.

Sleep

Most bird species are active during the day and sleep at night—except during migration, when the migratory flights of most species occur at night. Generally, a migratory bird begins a night's flight shortly after sunset and flies for about half the night. By migrating at night a red-eyed vireo experiences decreased predatory pressure, benefits from improved atmospheric conditions for flight, and is able to allocate more time to feeding during the day. All well and good, but nocturnal migration presents a time budget problem.

If birds normally sleep at night, then a nocturnal migrant experiences loss of sleep unless it compensates in some way. If loss of sleep has negative consequences for a migrating bird, migrants should show compensatory adjustments to sleep loss. Migratory birds might devote more time to sleep during the day, and brief periods of sleep have been observed during the late afternoon just before the onset of nightly migratory restlessness. Even if a migrant did rest during this quiescent period, though, the amount of sleep during migration would be less than the amount acquired during the nonmigratory season. Moreover, compensatory sleep during the day is not without costs for a migratory bird. Time available for other activities, including foraging, may be compromised, and the consequences, especially for individuals that were already energetically stressed after a long flight, could be serious. Little attention, if any, has been devoted either to sleep in migratory birds or to the possible consequences of sleep loss during migration.

Migrating birds are likely to solve en route problems if they settle in high-quality habitat, and the cheniers of coastal Louisiana are good places to stop over following a trans-Gulf flight. When my students and I compared the rate at which migrants replenished depleted fat stores at Peveto Beach Woods and East Ship Island (a barrier island off the coast of Mississippi), the difference was often dramatic (Fig. 4.3). White-eyed vireos (Plate 4.6) and black-and-white warblers are better off if they stop

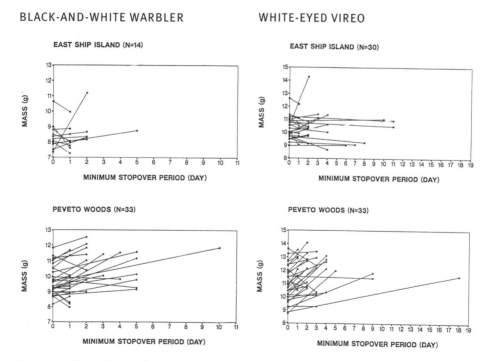

Figure 4.3 Change in mass between presumptive arrival and departure for (*A*) black-and-white warblers and (*B*) white-eyed vireos at two coastal woodland sites during spring migration, 1988.

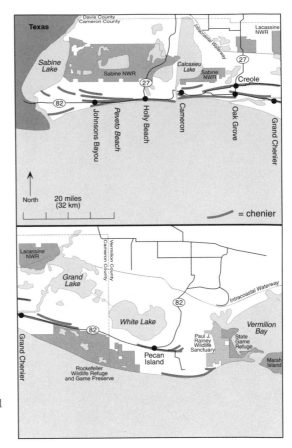

Map 4.1 Distribution and aerial
extent of oak chenier habitat in
coastal southwestern Louisiana

in the chenier, where they are more likely to gain mass in a hurry. Given the vicissitudes of weather, finding the right place is not an easy task for a small songbird after a trans-Gulf flight and it certainly involves an element of chance.

Moreover, coastal habitat has suffered degradation in the face of the increased development of coastal regions since World War II. Between 1960 and 1985, the human population living within 50 miles of the U.S. coast increased from 92.7 million people to 125 million people—52% of the population in the conterminous United States according to the U.S. Department of Commerce. The concentration of the U.S. population along our coasts is projected to continue well into the next century. By the year 2010, coastal populations will have grown to more than 127 million people, an increase of 60%.

The northern coast of the Gulf of Mexico, arguably the most important migratory stopover area for songbirds in North America, is expected to see significant human population increases. The southward migration of industry coupled with changing demographics will increase development pressure on stopover habitats in the decades ahead. Some coastal habitats spared from development are threatened by accelerating rates of erosion. The combined effects of coastal subsidence, the disruption of sediment supplies, and sea level rise will add further to the loss of important stopover habitats. Even today, the coastal cheniers are a minuscule collection of oak islands in a vast sea of marsh and prairie (Map 4.1).

As coastal areas are developed, there is a commensurate increase in the value of unaltered habitat to migratory birds. True, many migrants on most spring days pass over coastal woodlands and stop in inland forests [9, 10], but on those occasions when weather conditions are not favorable for continued migration [11] or when fat stores are all but gone, the isolated coastal woodlands and narrow barrier islands that lie scattered along the northern coast of the Gulf of Mexico become critical stopover habitat. Hence, preservation and restoration of these limited habitats and even the creation of new ones to replace those lost to coastal development should be a major conservation challenge in the twenty-first century.

References

1. Moore, F.R., and P. Kerlinger. 1987. Stopover and fat deposition by North American wood-warblers (Parulinae) following spring migration over the Gulf of Mexico. Oecologia 74:47–54.

   A technical paper describing the results of the author's studies on refueling during stopover.

2. Alerstam, T. 1990. Bird Migration. Cambridge: Cambridge Univ. Pr.

   An authoritative and somewhat technical, but readable, general treatment of bird migration; most examples are European.

3. Moore, F.R., S.A. Gauthreaux Jr., P. Kerlinger, and T. Simons. 1995. Habitat requirements during migration: important link in the conservation of Neotropical landbird migrants. *In* Ecology and Management of Neotropical Migratory Birds (T. Martin and D. Finch, eds.). New York: Oxford Univ. Pr.

   A general synthesis of information about the stopover ecology of landbird migrants; for the most part accessible to a general audience.

4. Moore, F.R., and T.R. Simons. 1992. Habitat suitability and the stopover ecology of Neotropical passerine migrants. In Ecology and Conservation of Neotropical Migrant Landbirds (J. Hagan and D.W. Johnston, eds.). Washington, DC: Smithsonian Inst. Pr.

   A synthesis of information on stopover ecology with an emphasis on habitat; readable by a general audience.

5. Blem, C. R. 1980. The energetics of migration. In Animal Migration, Orientation, and Navigation (S.A. Gauthreaux Jr., ed.). New York: Academic.

   A thorough technical review with an emphasis on birds.

6. Loria, D.E., and F.R. Moore. 1990. Energy demands of migration on red-eyed vireos, *Vireo olivaceus*. Behav. Ecol. 1:24–35.

   A technical paper containing original research data.

7. Moore, F.R., and W. Yong. 1991. Evidence of food-based competition during migratory stopover. Behav. Ecol. Sociobiol. 28:85–90.

   A technical paper containing original research data.

8. Kerlinger, P. 1989. Flight Strategies of Migrating Hawks. Chicago: Univ. Chicago Pr.

   Excellent treatment of raptor migration; technical, but general readers should find it informative.

9. Gauthreaux, S.A. Jr. 1971. A radar and direct visual study of passerine spring migration in southern Louisiana. Auk 88:343–365.

   A technical paper describing the results of the first intensive radar studies of spring trans-Gulf migration.

10. Gauthreaux, S.A Jr. 1972. Behavioral responses of migrating birds to daylight and darkness: a radar and direct visual study. Wilson Bull. 84:136–148.

    A technical paper describing the flight, flocking, and landing behavior of spring trans-Gulf migrants.

11. Buskirk, W.H. 1980. Influence of meteorological patterns and trans-Gulf migration on the calendars of latitudinal migrants. In Migrant Birds in the Neotropics (A. Keast and E. Morton, eds.). Washington, DC: Smithsonian Inst. Pr.

    A readable account of the role of broad-scale weather patterns on the occurrence and evolution of migration in the Gulf region.

5   JAMES BAIRD

*Returning
to the Tropics:
The Epic
Autumn Flight
of the Blackpoll
Warbler*

It is difficult for the beginner in bird study to believe that the little green birds that come trooping down from the north in autumn are the same that went north in their black and white vesture in the spring; silently they pass or with only a faint lisping chirp in place of the songs of the spring migration. From far-off Alaska, from the great Northwest Territories, from Hudson Bay and Labrador, the Black-polls come down toward the peninsula of Florida and steer their course across Cuba and the West Indies to South America.
—Edward Howe Forbush, *Birds of Massachusetts and Other New England States,* 1929

When those words of Forbush were published, in the last year of his life, I am sure that he had no doubt as to their veracity, despite the complete absence of any relevant empirical evidence. The spring migration of the blackpoll warbler, northward through Florida and then fanning out to reach the breadth of the North American boreal forests, was well known. It must have seemed beyond question that they retraced that route during the autumn return to the tropics. Only recently did ornithologists begin to put together the disparate pieces from which emerged a very different picture of

what the blackpoll actually does in the fall. The transoceanic flight that Jim
Baird describes must be one of the greatest tests of endurance engaged in by
any animal. The distance covered translates to about 35 million body lengths,
not as long, relative to body size, as that traversed by the rufous hummingbird (see Chapter 10), but performed in a single, nonstop marathon flight. It
is not yet possible to track an individual blackpoll warbler from New England to Venezuela; so as Jim makes clear, our belief that they do so is based
on a collection of different types of evidence (distribution of migrants, body
weights recorded at various places, radar surveillance, data collected from
birds killed during migration). The case is necessarily a circumstantial one,
involving some nifty ornithological detective work. Most, but not all, ornithologists today find the scenario he describes a compelling one. Doubters
in science are important, for they cause us to continually question our conclusions and retest our hypotheses, and that is the process which moves us
inexorably closer to the truth. Had Forbush been confronted with skeptics
questioning his comfortable assumptions, he might have avoided perpetuat-
ing the erroneous story of the blackpoll's autumn migration back in 1929.
—K.P.A.

It was a flight of angels seemingly bent on suicide that piqued the curiosity of two
scientists and launched a continuum of studies and reports that have spanned nearly
four decades. For it was in 1961 that William H. Drury and J. Anthony Keith began
their analysis of films from a radar site located on Cape Cod, Massachusetts, the
most puzzling aspect of which was the immense departures at night of radar angels
whose heading was southeast. This was enigmatic, for while Drury and Keith were
certain that the radar echoes were indeed birds and, given their flight characteristics,
were most likely songbirds, they were aware that only the inhospitable western Atlantic lay southeast of Cape Cod. Unless, of course, they wanted to believe that these
angels intended to fly nonstop to the West Indies or to the northern coast of South
America, an improbable 1,600 to 2,200 miles distant.

However, secure in the knowledge that events of this magnitude were unlikely to
be the result of natural selection gone awry, Drury and Keith began to look for an
explanation of this phenomenon. First they made a systematic review of the possible
avian candidates that might benefit from such a venture, finding that there were 13
species of wood-warblers that occurred in the northeast and wintered in the West Indies and South America. It was soon determined that, of this group, the blackpoll
warbler merited further investigation because a) it was abundant, b) it was known to
store the ample deposits of fat that might fuel such a long flight, and c) it wintered

in South America. And so it was that with the emergence of such a strong candidate, but not excluding the other 12, Drury and Keith proposed in 1962 the hypothesis that significant numbers of songbirds migrated from New England and Nova Scotia across the western Atlantic to wintering grounds in the West Indies and South America [1]. This study was followed in 1963 by a two-part paper by Nisbet, Drury and Baird [2] that focused solely on the blackpoll warbler: part I was a detailed analysis of data concerning the autumn migration of the blackpoll warbler and was specifically directed at determining whether blackpolls could store enough energy (fat) to power a flight of 2,200 miles; part II was a review of other estimates of flight energetics by Nisbet. Since then, there have been 25 papers generally supporting this hypothesis and four papers, by one author, in persistent opposition [3].

Today, while the mounting evidence remains circumstantial, the hypothesis is generally accepted by ornithologists that the blackpoll warbler (and possibly other species) migrates in the autumn from the northeastern United States across the western Atlantic to South America. Here I want to present to you the scientific detective story that brought together the disparate pieces of evidence that collectively tell a tale of remarkable adaptation and endurance.

Like all other wood-warblers, the blackpoll warbler is small, measuring a scant 4 inches in length and weighing when lean between 9 and 11 g (which is about the weight of a ballpoint pen). The breeding plumage is acquired by a partial molt toward the end of the birds' stay in the tropical forests of northern South America and is completed by the time that they begin migrating to their northern breeding grounds. The male is now a strikingly patterned black and white (Plate 5.1), while the female is a black-striped olive gray. Blackpolls begin their northward trek in late March and early April when they pass through the West Indies to Florida and to the eastern Gulf Coast of the United States. From here they fan out across the eastern United States, arriving at midcontinent by the middle of May and at their boreal-forest nesting grounds in early June. Some take longer than others to get "home" because this species has an immense breeding range that stretches across the width of the continent from Alaska to Newfoundland, extending north to the tree line and south to southern Quebec and the mountains of New England and New York (Map 5.1).

At the latitude at which they nest, the summer is short but the days are long. By the middle of July, nesting is essentially completed and both adults and young begin their molt—a partial molt for the young and a complete one for the adults. The end result is a remarkably homogeneous population of adults and young that are black-striped olive-green birds. By mid-August when the molting period is over, the whole population starts to converge on the northeastern United States for the first leg of their incredible journey.

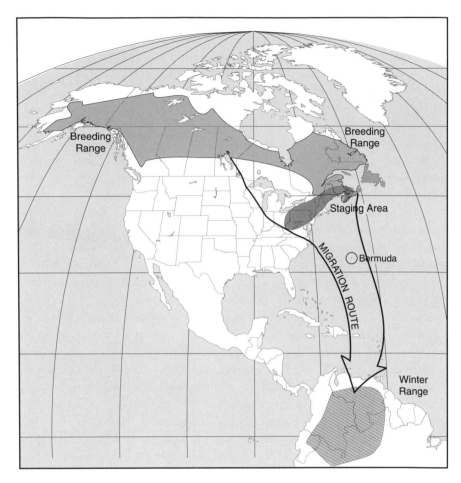

Map 5.1 The breeding range, wintering area, migration route, and preflight staging area of the blackpoll warbler.

The Overland Flight

As raindrops on their way to the sea, the postbreeding family groups of blackpolls gradually coalesce into small flocks, then into larger flocks which independently flow from the nether regions of their breeding grounds toward their first destination: the staging area that encompasses New England south to Virginia and west to West Virginia and western Pennsylvania (Map 5.1). Some, like the eastern populations, have to move a relatively short distance, 600 to 800 miles, while those from the western Canadian provinces and Alaska travel up to 1,500 miles.

This is an orderly exodus and it follows traditional routes with well-defined boundaries: hence, blackpolls are common or even abundant as migrants north of a line that extends from Minnesota to West Virginia, but they are uncommon, rare or totally absent south of that line. For example, blackpolls in the fall are regularly common in the Chicago area but are less predictable in central Illinois, where their appearance appears to be correlated with exceptionally strong cold fronts. In southern Illinois they are uncommon. Similarly, hundreds, sometimes more than a thousand blackpolls are banded each year at the Allegheny Front Migration Observatory in the mountains of West Virginia (Plate 5.2), while they are extremely rare in the bordering state of Kentucky.

This flight is not without peril, for like other passerine migrants the blackpoll warbler is subject to natural hazards. Increasingly it is man-made objects that pose a threat against which birds have no defense. Tall buildings, lighthouses, and television towers are the greatest obstacle to safe passage [4]. No hazard during fair weather, the lights on these structures have a fatal attraction to passing nocturnal migrants during periods of cloudiness, fog, or rain. Once drawn toward the structure's lights, they mill about in vocal confusion with their calls attracting other birds. Eventually they start colliding with the building, the tower, or the tower's guy wires in a destructive accretion that routinely results in the death of hundreds and even thousands of migrants in a single night at a single location. Given the ubiquity of tall buildings and television towers, the annual toll across the continent is immense.

Regrettable as these misfortunes may be, they are nevertheless a boon for ornithologists, for the kills provide an otherwise unobtainable sampling of birds plucked from the night sky while migrating. Each event provides the investigator with data about the species composition of the flight, as well as information about the age and sex, weight, wing lengths, and other physical data for each species. And it is the death of all these birds over a span of years that will, in the end, provide us with some of the information needed to understand better the complexities of bird migration, thereby enabling us to plan for their future management and ultimate survival.

As an example of how science has benefited from these tragedies, the large sampling of blackpoll warblers killed in transit from midcontinent to the Atlantic, when combined with other data, provides us with insight into this first stage of their autumn migration. I have personally examined more than 1,700 blackpolls killed at television towers, lighthouses, or other lighted structures. These data have told us the following: a) blackpolls are on the move shortly after they complete their molt (but not before, since no evidence of molt was noted in any of the birds examined, all of which were freshly plumaged); b) the first migrants show up on the periphery of the staging area in late August and early September, the largest numbers arrive throughout September, and the late arrivals continue until mid-October; c) black-polls do not travel in discrete sex or age groups and, as might be expected at the close of the breeding season, immatures outnumber adults by a 2:1 ratio; d) the average weight of these overland transients is 12.5 g. Because it has been determined that the fat-free weight of blackpolls is between 9.3 and 11.3 g [2], it is obvious that for most of the birds (<75%), this first overland stage of their migration is undertaken with small amounts of fat. This fact has been noted in other species that engage in "premigratory" or "intramigratory" short hops [5]. While specific data on the distance covered by blackpolls in a single night are not available, it seems reasonable to assume that, given the amount of fat available, a ground speed of 27 mph [6], and a flight of 5 to 8 hours, they migrate between 150 and 250 miles per night.

The Staging Area

With the passage of each cold front from early September to mid-October, black-polls stream into the northeastern part of the eastern deciduous forest, where they settle in to rest, feed, and fatten up for the next and more arduous part of their journey. As diverse as the landscape upon which it rests, this ecosystem is composed of a mixed forest of ashes, maples, oaks, and birches and it is interspersed with an abundance of white pine and other conifers, depending upon the site. Typically, in Massachusetts, for example, after a night's flight, the blackpolls land in the predawn hours and, possibly, sleep briefly. Then, starting about a hour after dawn, large numbers of blackpolls engage in a low-level (100–500 ft) movement to the west, northwest, or north that lasts from 1 to 4 hours. It is a broad-front movement, most visible from prominences, with the direction apparently but not always influenced by the prevailing wind: these early-morning flights tend to be made into the wind. Some idea of the numbers involved in these flights can be gleaned from a sampling of observations that I made in Littleton, Massachusetts, where on 12 September 1967, blackpolls were passing overhead at the rate of 250 per hour and where, on the same date a year later, 749 were seen from 8:20 to 10:00.

While it is possible that this phenomenon is rooted in the same innate behavior of other migrants along the coast — viz., to get away from the coast — such an explanation seems unlikely inasmuch as these movements occur well inland. A more reasonable hypothesis might be that this morning flight is an effort by the participants, after landing in the dark of night, to return to familiar territory, having possibly overshot their intended destination [7]. A strong argument in favor of this proposal is found in the demonstrated ability of blackpolls in Massachusetts, and presumably elsewhere, to return to the same site where they had been banded in the previous year. This most certainly was the case with 3 banded birds in 1963 and 1 in 1968 (Plate

5.3). While the number of returning birds is small (4 out of 2,695 banded), the odds against these being chance happenings are so astronomical that we must conclude that each of these birds was by choice at the same location where it had been the previous year. Equally astonishing is their timing; in two of the four instances, the birds returned within 1 day of the date they were banded, another within two days, and the fourth fifteen days later [8]. Even if this homing ability is unrelated to the morning flight, it is nonetheless remarkable. That birds have a well-developed sense of place is well known, as exemplified recently by site-fidelity studies of American redstarts and black-throated blue warblers that showed return rates, respectively, of 27% and 37% to their breeding grounds and 51% and 46% to their wintering sites [9]. And it has been well documented that many shorebirds are heavily dependent upon traditional staging areas for the successful completion of their migration (see Chapters 8 and 9). So it is by no means improbable that blackpoll warblers have the ability to find and to make use of a specific location midway between their nesting and wintering grounds.

Wave after wave of blackpolls enters the staging area, and whether it is Massachusetts, New York, Pennsylvania, or West Virginia, they arrive at their intended destination and join together in small flocks of 10 to 40 individuals that are often augmented by small numbers of other warbler species. The flocks move ceaselessly about in search of good food sources, and occasionally an arthropod bonanza is struck. This happened at Round Hill in Sudbury, Massachusetts, in the fall of 1964, when aphids infested the gray birches that capped the hill. The number of aphids was incredible, since virtually every leaf of every tree was laden with 3 to 10, occasionally as many as 22, aphids. But while episodic food troves are undoubtedly capitalized upon when located, it seems more likely that what these flocks are seeking is an area that can provide a consistently rich food source and to which they will return in subsequent years.

Gradually the blackpolls lay on ever-increasing amounts of fat, as is apparent when the mean weight of the birds on the staging area is compared with their weights in transit. However, even more significant than the mean weight of the staging-area birds, which is generally depressed by the low weights of new arrivals that continue well into October, is the high percentage of birds that weigh more than 13.5 g. By way of explanation: birds are "scored" on their degree of fatness, with fat classes ranging from 0 (no fat) to 5 (very fat); the 13.5-g figure is the average weight of 240 birds in fat class 2, thus providing a rough measure of the increasing fattiness of the birds in the staging area. But while the weight of these populations as a whole appears to be increasing gradually, individuals are often, in fact, putting on weight very rapidly, the consequence of *hyperphagia*—an activity that might best be described as gluttony with a purpose. It should be noted that this activity is not unique to blackpolls, for many migrant songbirds accumulate large amounts of fat prior to or during migration [10]. In blackpolls, however, hyperphagia can produce dramatic increases in weight in a short time, with many birds weighing more than 20 g (the heaviest bird recorded to date weighed 25.6 g on 30 September 1964). Not all individuals achieve weights of 20 g or more, though, and it appears that the departure weight of blackpolls setting out to cross the ocean can vary by as much as 10 g.

The Flight across the Western Atlantic

Weather plays an important role in the successful completion of the blackpoll's odyssey. The weather regime in North America in the fall is such that masses of cold air periodically emanate from northern Canada and flow across the eastern United States. The leading edge of this high-pressure cell, the *cold front*, moves forward at varying rates of speed, depending upon the dynamics of the air masses, until it crosses the coast, passes out over the Atlantic, and eventually stalls a few hundred to eight hundred miles offshore. Typically, the passage of a cold front brings with it a drop in temperature, clear skies, and northwest winds.

As was the case with the blackpolls entering the staging area, it is the passage of a cold front that stimulates those physiologically ready birds to launch themselves into the air that night, set their course to the southeast, and begin their journey over the open ocean. Flying at an air speed of more than 20 mph at an altitude initially between 1,500 and 2,500 feet, the migrants are assisted by the northerly air flow that follows behind the cold front. Sometimes the wind behind the front can be quite strong and other times less so, but even a light following wind of 6 to 10 mph will increase the birds' ground speed by one-third to one-half. Gradually the migrants merge into flocks numbering 500 to 1,000 birds. Eventually they catch up with and pass through the stalled front, and by the second day they are in the relatively still air at the western edge of the Sargasso Sea. Hour after hour they fly, now at an altitude of between 3,500 and 7,000 feet; by the third day they are well within the zone of the northeast trade winds, where once again the following winds boost their flight speed. As they draw closer to the Lesser Antilles, the winds at lower elevations become more easterly and stronger, representing more of a hazard than a benefit, so the birds increase their altitude to 15,000 feet where the air is calmer [11]. As the blackpolls approach the South American mainland, they drop lower and lower and land after a nonstop flight of 2,200 miles.

That, in a nutshell, is what this hypothesis is all about, but what evidence is there to support it? First, there are the observations with radar. Since 1962, there have been eighteen papers reporting on studies that used radar to investigate one or another aspect of bird migration over the western Atlantic. Most were studies from a single location, which, given the 1- to 110-mile range limitation of the search and/or the narrow-beam radars used, meant that each analysis was pertinent to a rather limited area around the radar site. However, when considered in the aggregate, these disparate studies led to a reaffirmation that large numbers of songbirds were migrating in a purposeful manner over the ocean.

But it wasn't until Timothy C. Williams and his colleagues embarked on a 6-year study employing ships at sea plus using a radar network of eight sites—at Halifax, Nova Scotia; Cape Cod, Massachusetts; Bermuda; Wallops Island, Virginia; Miami, Florida; Antigua; Barbados; and Tobago—that waves of transoceanic migrants were tracked from departure to destination. The Williams team found that the large numbers of songbirds departing from the North American coast first headed south-southeast, passed over Bermuda heading southeast, and then passed south-southwest over the Lesser Antilles. And, as already noted, they also determined that the birds flew at a ground speed of about 27 mph at an altitude of between 3,500 and 7,000 feet, and that this flight is accomplished in 82 to 88 hours [6, 11].

As useful as these radar studies were, it still remained to be determined what species were involved. Fortunately, Bermuda is well positioned to provide exactly the kinds of data needed. Located about 750 miles east-southeast of Cape Hatteras, North Carolina, Bermuda lies well within the flight path of transoceanic migrants, as is well documented by radar [6]. Paradoxically, if all was going well for all these migrants passing over, Bermuda's position would be of little benefit since they would continue on their way unseen, as they do most of the time. But this is not a perfect world, and there are three sets of data from Bermuda that were dependent upon calamity.

As is so often the case with islands, Bermuda offers a haven for thousands of lost waifs, birds that were not meant to be wandering around in the vastness of the Atlantic Ocean. Some idea of the magnitude of this phenomenon is evident in the fact that 87 species of North American passerine migrants are seen regularly on Bermuda each autumn [12]. Most of these birds are immatures (Plate 5.4). This is no surprise, for it is well documented that many immature birds 1) are poor navigators and/or prone to drifting off course and 2) tend to become concentrated along the coast in the fall or to become lost at sea [13, 14]. To illustrate the point: of 266 fall migrant passerines banded in Bermuda from 1963 to 1965, often at very low weights, 256 (96%) were immatures [D.B. Wingate, letter to author]. Despite the birds' having found brief respite on Bermuda's islands (although a few do stay to winter), their chances of getting back into the gene pool are slim. These birds are, in essence, culls in a selective process that impacts every migratory species in one way or another. Deliberately excluded from this count were 166 blackpoll warblers that were banded during the same period, because it was obvious from the 1:1 age ratio (83 adults and

83 immatures) that their presence on Bermuda was based on a different set of circumstances than that of the other species.

Data gathered at the staging area suggested that blackpolls departed on their transoceanic journey at a wide range of weights. For example, one adult and six immature birds killed at night in Boston, on 22 October 1969, had weights ranging from 15.2 to 20.8 g. A group of blackpolls killed at sea 43 miles south of the Massachusetts coast, on the night of 9 October 1979, was evaluated on the basis of fat class [15]. It should be noted that fat-scoring can be a rather precise method of indexing total body fat when performed by an experienced investigator [16]. Of the forty-four birds, three were scored as fat class 1, four as fat class 2, eleven as fat class 3, and twenty-six as fat class 4. Using blackpoll data from the Manomet Observatory, the authors found that fat class 3 was equivalent to an average weight of 16.5 g and that fat class 4 equaled 19.6 g. Thus, although a considerable range of body weights was represented in this sample of departing migrants, the great majority were heavy. A sample of 216 blackpolls killed while migrating past Bermuda over a period of years (1959–1968) also showed a wide weight range, from 13.5 to 22.1 g.

The importance of this weight range is apparent after consideration of the following: 1) blackpolls utilize fat during migration at the rate of about 0.08 g/h [2, 17]; 2) the distance between Massachusetts and Bermuda is approximately 800 miles; 3) migrating at a ground speed of 27 mph, blackpolls would take about 30 hours to reach the latitude of Bermuda, thus utilizing 2.4 to 2.6 g of fat. Those birds that depart without sufficient fat are in serious trouble by the time they reach Bermuda, and it is these birds (both adults and immatures) that for the most part populate Bermuda every autumn. Other blackpolls may seek shelter on Bermuda because of other factors, such as bad weather, which leads to a somewhat incongruous situation: newly arrived blackpolls have been captured in the same net at the same time, one weighing 8 g. and the other 18 g. However, the lucky blackpolls that make it to Bermuda are but a small fraction of the numbers involved, and they can and do put on weight and continue on their way; the others, out of sight of Bermuda, perish at sea and leave the legacy of a strengthened gene pool.

Occasionally blackpolls encounter storms or generally inclement weather that forces them to fly much lower than is usual, almost at sea level. This low altitude, coupled with the fatal attraction that lights have for blackpolls during foul weather, causes them to be drawn to or killed at the Gibbs Hill Lighthouse on Bermuda or to the Argus Tower located 24 miles southwest of the island. (The tower is a former early-warning platform structure, installed and operated by the United States and, at the time of these observations, devoted to naval research.) Birds captured, banded, and released at the Gibbs Hill Lighthouse in 1959, 1960, and 1962 were flying over Bermuda with no apparent intention of stopping. Their weights provided the keystone in the weight-loss analysis, offering reasonable proof that the blackpolls could indeed make the transatlantic flight [2]. The large sample of 185 dead birds from the Argus Tower in 1967 and the smaller sample of 13 from 1968 proved, if such were needed, that the 1962 event wasn't an anomaly. The importance of these data was recognized from the outset.

Taken as a whole, these data, gathered over a period of 10 years, show that virtually all of the blackpolls passing over Bermuda at night have ample and sufficient fat reserves to fuel a flight of an additional 1,350 miles. The lowest-weight bird in 1967 weighed 14.3 g. Given that it would take an additional 50 hours for a blackpoll to fly to South America and using the fat utilization factor (0.08) noted above, it can be estimated that it takes 4 g of fat to fuel this flight. So the bird in question would arrive weighing 10.5 g, well within the normal weight range of a nonfat blackpoll. Even birds weighing 13.5 g could make the trip successfully, although it would have to be an uneventful trip. All those birds weighing 18, 19, or 20 g or more by the time they reach Bermuda are clearly capable of a longer sustained flight. Evolutionarily speaking, these heavier birds are provided with a substantial margin of error if they should run into a tropical storm or hurricane that would require remaining aloft longer than usual.

It is significant that in all the reports of nocturnal accidents from ships or on Bermuda that, with one exception, only the blackpoll warbler has been involved. That exception is the Connecticut warbler [18], an uncommon species that nests in spruce-tamarack and aspen-poplar forests in the north-central United States and central Canada and that winters in north-central South America (Plate 5.5). Like the blackpoll, Connecticuts become very fat, are rare in the southeastern United States, but are regular fall migrants in the northeastern United States. In spring they migrate across the Gulf of Mexico and up the Mississippi Valley to their breeding grounds. On 6 October 1967, David B. Wingate observed an enormous flight of birds passing by (and also striking) the Argus Tower southwest of Bermuda. He was able to identify, as they passed through the lights, not only blackpolls (which were the most common) but also Connecticuts (which were second in abundance). Other species were rare, but those that could be positively identified included a magnolia warbler and a female common yellowthroat—only a handful out of the thousands passing by [D.B. Wingate, letter to author, 28 Nov. 1967].

In summary, like all biological entities, blackpoll warblers are a work in progress. One facet of this work, and a critically important one, is their autumn migration. There is a distinct advantage for this species to fly across the western Atlantic: it is about 1,500 miles shorter than a route via Florida, thereby allowing blackpolls to go from the staging area to the wintering area in the shortest time possible. The crossing is, with rare exceptions, predator-free and, in the long view, relatively free of catastrophic meteorological events; most of the individuals are capable of making this flight with ease. There is, however, an unknown number that cannot make the flight because of insufficient fat storage. Their loss at sea will, in the end, leave the surviving population better adapted to make its epic flight.

The Alternative Hypothesis

As noted, there is another view of how the blackpoll warbler gets from its breeding ground to its wintering ground. Bertram G. Murray Jr., in a series of papers, has taken the position that "blackpoll warblers funnel from the breeding range to the southeastern United States, along with the general movement of migration . . . and depart from somewhere between Cape Hatteras and northern Florida." [3, 14, 19, 20]. The basic problem with his hypothesis is that blackpoll warblers are uncommon to rare autumn migrants south of Cape Hatteras, North Carolina, whereas north of Hatteras, they are common [21].

Epilogue

It was a bright early-October morning in Bermuda. We were all lounging about in the deck chairs scattered around the pool at the White Sands Hotel, which had been our headquarters during our weeklong Massachusetts Audubon tour of Bermuda. David Wingate, Bermuda's conservation officer and our guide, had just said his good-byes and left us to head back to his office, while we waited for taxis to take us to the airport. It had been a good trip. There had been lots of birds, good snorkeling, plenty of beach and pool time, and our coverage of Bermuda was exhaustive—Spittal Pond, Ferry Reach in St. Georges, the Arboretum, the Hamilton dump, and Wingate's version of Eden, Nonsuch Island. So it was a contented group relaxing in the balmy 70-degree temperature, made all the sweeter by the knowledge that it would be quite some time before we experienced such temperatures again. However, relaxing doesn't come easily for me when there are birds to be looked at, and so I spent most of my time searching the ocean, the trees, and the sky for that last special bird. The pickings were slim, and I was sometimes reduced, happily enough, to watching the interaction between two active and very noisy kiskadees and even the warily brazen house sparrows that hopped under the poolside tables in search of edibles. When suddenly and unbelievably I spotted a flock of birds high overhead. Binoculars up, focused, and Holy Smokes! There was a flock of 23 great blue herons! They were barely visible to the naked eye and it was clear that Bermuda was not on their itinerary as they flapped steadily onward.

I was really excited. I had seen great blue herons both leaving and arriving at the south coast of New England, but I had always assumed they were taking a shortcut across to Long Island or New Jersey. I never dreamt that they might have just crossed or were about to cross the western Atlantic. But here we were, 750 miles from the nearest land, and a flock of great blue herons was passing overhead on what appeared to be a purposeful southward flight. These were not lost birds. Wow! I was impressed.

Ever the diligent tour leader, I quickly set up the telescope so that members of the tour group could gaze upon this wonder. Most found the event underwhelming, so I had virtually unlimited access to the scope. And it was this fortuitous circumstance that led to an even more remarkable discovery: while I was watching the herons, there came into view, high above them, a flock of more than 1,000 passerines! This fairly tight flock, roughly spherical in shape, quickly left the herons behind. I followed them with the scope until they were out of sight.

My mind raced. What could they be? What species flocked like that? Thrushes? No. Finches? Well, could be. Finches did fly in flocks, but I had never seen a flock of finches quite like this—they tended to be looser. Not only that, but what would they be doing here? They wintered for the most part in North America, and these birds acted like they knew where they were going—and it wasn't Georgia or Florida. Bobolinks! That's it, I thought, they must be bobolinks. They fly in flocks and they winter in South America. Incredible! I couldn't believe my luck. I reveled in the thought that I had just witnessed a fantastically rare event; not rare in its occurrence, but rare in its observation. For someone as interested in migration as I was, this was awe-inspiring and a fitting end to a trip that had focused largely on birds whose presence on Bermuda was the result of migratory misadventures.

So bobolinks it was in my memory until a decade or so later when Timothy Williams and his band of radar-watchers reported that the birds they had tracked, which presumably included blackpoll warblers, flew in flocks of 1,000 or more birds! Well now, this put a different light on my earlier observation: possibly, or dare I say probably, the birds that I had seen were blackpoll warblers. This was a possibility that hadn't crossed my mind at the time, but the new evidence makes it plausible.

Bobolinks or blackpolls? It mattered not. What was significant about the observation was that it proved to me that the concept of a voluntary migratory flight across the western Atlantic was valid, for here was a large flock of small birds seen flying high over Bermuda that showed no sign of being lost. Indeed, their altitude, their speed, the flocking behavior, and their steady course provided solid evidence that they knew where they were going.

Bobolinks or blackpolls. I'll never know for sure, but down deep I will always think of them as blackpolls. And I will feel privileged to have witnessed this extraordinary journey.

References

1. Drury, W.H., and J.A. Keith. 1962. Radar studies of songbird migration in eastern New England. Ibis 104:449–489.

    A classic early radar study of bird migration in North America; readable.

2. Nisbet, I.C.T., W.H. Drury, and J. Baird. 1963. Weight-loss during migration. Part I: Deposition and consumption of fat by the blackpoll warbler *Dendroica striata;* Part II: Review of other estimates by I.C.T. Nisbet. Bird-Banding 34:107–138.

    A technical paper that examines the fat deposition and potential flight range of the blackpoll warbler.

3. Murray, B.G. Jr. 1989. A critical review of the transoceanic migration of the blackpoll warbler. Auk 106:8–17.

    A critique of the hypothesis of the long overwater flight and an espousal of the alternative hypothesis.

4. Nisbet, I.C.T. 1970. Autumn migration of the blackpoll warbler: evidence for long flight provided by regional survey. Bird-Banding 41:207–240.

    An enormous compilation of distributional data from which Nisbet inferred that blackpolls must be making a transoceanic flight.

5. Johnston, D.W. 1966. A review of the vernal fat deposition picture in overland migrant birds. Bird-Banding 37:172–183.

    A classic early study of spring fat deposition in passerine migrants.

6. Williams, T.C., J.M. Williams, L.C. Ireland, and J.M. Teal. 1978. Estimated flight time for transatlantic autumnal migrants. Am. Birds 32:275–280.

    A readable account based on radar studies from New England to the Caribbean.

7. Gauthreaux, S.A. Jr. 1978. Importance of daytime flights of nocturnal migrants: redetermined migration following displacement. In Animal Migration, Navigation and Homing (K. Schmidt-Koenig and W.T. Keeton, eds.). Heidelberg: Springer-Verlag.

    A visual study of the early-morning flights of predominantly nocturnal migrants; concludes that the birds are making a correction for wind drift experienced during the night.

8. Nisbet, I.C.T. 1969. Return of transients: results of an inquiry. EBBA News 32:269–274.

    An examination of band recoveries showing that migrants sometimes return to stopover localities used on previous migrations.

9. Holmes, R.T., and T.W. Sherry. 1989. Site fidelity of migratory warblers in temperate breeding and Neotropical wintering areas: implications for population dynamics, habitat selection, and conservation. In Ecology and Conservation of Neotropical Migrant Landbirds (J.M. Hagan III and D.W. Johnston, eds.). Washington, DC: Smithsonian Inst. Pr.

    A technical paper that looks at the issue of site fidelity in relation to breeding as well as overwintering sites.

10. Blem, C.R. 1980. The energetics of migration. In Animal Migration, Orientation and Navigation (S.A. Gauthreaux Jr., ed). New York: Academic.

    A detailed and technical review of fattening and the energetics of migration in birds and other animals.

11. Williams, T.C., J.M. Williams, L.C. Ireland, and J.M. Teal. 1977. Autumnal bird migration over the western North Atlantic Ocean. Am. Birds 3:251–267.

    A readable account of radar studies spanning the whole transoceanic route.

12. Wingate, D.B. 1973. A Checklist and Guide to the Birds of Bermuda. Hamilton, Bermuda: Island.

13. Baird, J., and I.C.T. Nisbet. 1960. Northward fall migration on the Atlantic coast and its relation to offshore drift. Auk 82:119–149.

    A classic study based on observations of northwestward movements of fall migrants along the New England coast and offshore islands.

14. Murray, B.G. Jr. 1966. Migration of age and sex classes of passerines on the Atlantic coast in autumn. Auk 88:352–360.

    A review of data from coastal banding stations; documents the overwhelming preponderance of young of the year among migrants on the coast.

15. Cherry, J.D., D.H. Doherty, and K.D. Powers. 1985. An offshore nocturnal observation of migrating blackpoll warblers. Condor 87:548–549.

    Analysis of the condition of blackpoll warblers that came aboard a ship during nocturnal migration.

16. Krementz, D.G., and G.W. Pendleton. 1990. Fat scoring: sources of variability. Condor 92:500–507.

    An evaluation of the most commonly used method of assessing the fat loads of migrants.

17. Greenewalt, C.H. 1975. The Flight of Birds. Trans. Amer. Philos. Soc. 65(4).

    Classic introduction to bird flight.

18. Curson, J., D. Quinn, and D. Beadle. 1994. The Warblers of the Americas. Boston: Houghton Mifflin.

    A well-illustrated general account of the life histories of the wood-warblers.

19. Murray, B.G. Jr. 1965. On the autumn migration of the blackpoll warbler. Wilson Bull. 77:122–133.

    The original formulation of Murray's hypothesis that blackpolls do not make the long overwater flight.

20. Murray, B.G. Jr. 1966. Blackpoll warbler migration in Michigan. Jack-pine Warbler 44:23–29.

    Discusses blackpoll migration in Michigan in relation to Murray's contention that the birds do not make the transoceanic flight.

21. McNair, D.B., and W. Post. 1993. Autumn migration route of blackpoll warbler: evidence from southeastern North America. J. Field Ornithol. 64:417–427.

    Data on the scarcity of blackpolls in autumn in the southeastern United States, where Murray proposes they stage for a shorter overwater flight.

KEITH L. BILDSTEIN

Doth the hawk fly by thy wisdom, and stretch her wings toward the south?
—Job 39:26

Racing with the Sun: The Forced Migration of the Broad-winged Hawk

I first met the broad-winged hawk as a child in the forests of southern Kentucky. I saw it only occasionally during the breeding season, usually as it slipped quietly from a perch beneath the forest canopy and disappeared into the trees. Like the early naturalists, I knew nothing of its extraordinary migration and assumed, in my ignorance, that it lived in the area all year. Since then I have encountered it at many places on its appointed track. When I lived near the Great Lakes, the arrival of the first push of broad-wings was one of the surest signs of real spring along the south shore of Lake Erie. In high school I occasionally cut classes when my friends and I predicted a big flight based on our consultation with the weather map. Sometimes we were right, and on those days I learned more than I could have in school. A few years ago, in Costa Rica, we watched broad-wings, Swainson's hawks, and turkey vultures stream over the rain forests of La Selva, a seemingly endless procession gliding northward on motionless wings. There, and in the Northeast in both spring and fall, the hawks are seen primarily in the air. At night they disappear into the large forested areas, for the most part unseen until they arise magically on the next day's thermals. One spring in southern Texas, however, I observed a different facet of the broad-wing's migratory life. Enormous flights were passing over the lower Rio Grande Valley and into the cattle country to the north. Here, woodlands are small and isolated; yet the hawks must put down somewhere when the changes in the atmosphere dictate an end to the day's migration. Early one morning I was birding in some small oak mottes north of the valley. The short trees were literally full of broad-winged hawks, hundreds of them: it was like a blackbird roost. In a couple of hours they were on their way, but I was struck with the importance of these little islands of oak for transient hawks. Even though any one individual broad-wing will probably never spend another night here, for a few days each spring these small woodlands fill up every night like motels along the interstate, providing refuge and rest in an area where the migration stream is particularly constricted.

Among our North American raptors, the Swainson's and broad-winged hawks, osprey, peregrine falcon, Mississippi kite, and turkey vulture are the longest-distance migrants. Performing these journeys places significant time constraints on the birds, and wedging all of the necessary events of the annual cycle into a twelve-month period necessitates a number of evolutionary compromises. Keith Bildstein weaves all of the components together in this chapter, and some of the same themes will run through Chapters 8 and 9 dealing with shorebird migration. —*K.P.A.*

It is 9:00 A.M., 16 September 1995, Hawk Mountain Sanctuary. Thousands of people from the northeastern United States—others from as far away as Germany, France, and Japan—are streaming into a series of misshapen, gravel parking lots atop a windswept Appalachian ridge 25 miles northwest of Allentown, Pennsylvania. A group of orange-vested, radio-toting volunteers waves on the line of cars. The "vols" are doing their best to unsnarl the rural traffic jam, but Hawk Mountain Sanctuary is in the middle of its crazy season, and controlled pandemonium rules. Many of the impatient drivers have traveled long distances in anticipation of Hawk Mountain's main event, and nothing—including the car in front of them—is going to stop them short of their goal.

Hawk Mountain is venerated ground this time of year. Most of the faithful have come to see the en masse migration of the broad-winged hawk.

The North Lookout (Plate 6.1), the sanctuary's official count site, is three-quarters of a mile up the trail from the parking lots, and the walk is hurried. Nobody is saying so, but "Out of my way, buddy, I'm here to see hawks" fills the air.

Table 6.1 *Annual average counts of migrating raptors at Hawk Mountain Sanctuary, 1934–1995*

Species	Annual average
Turkey vulture[a]	143
Black vulture[a]	39
Osprey	342
Bald eagle	47
Northern harrier	223
Sharp-shinned hawk	4,246
Cooper's hawk	283
Northern goshawk	69
Red-shouldered hawk	245
Broad-winged hawk	8,527
Red-tailed hawk	3,208
Rough-legged hawk	9
Golden eagle	45
American kestrel	367
Merlin	33
Peregrine falcon	23
All raptors	17,787

[a]Data for turkey and black vultures are based on counts since 1990.

Founded in the summer of 1934 by the conservationist Rosalie Edge, Hawk Mountain Sanctuary is the world's first refuge for birds of prey. Sanctuary personnel have been recording the movements of migrating raptors at the North Lookout for more than 60 years [1–3]. During that time, more than a million raptors, representing 18 of the continent's 35 species, have been recorded flying past the site. By far, Hawk Mountain's most numerous migrant is a continentally endemic, crow-sized buteo known as the broad-winged hawk (Table 6.1). Hawk Mountain Sanctuary records an average of more than 8,000 broad-wings a year [4]. In most years, the broadwing flight peaks within a day or two of the 16th of September, which explains the crowd this morning.

An exhilarated Doug Wood, the sanctuary's official volunteer counter for the day, had arrived at the North Lookout at 5:30 A.M. Doug was an hour early, in eager anticipation of a great flight. A cold front, the synoptic weather event that Sanctuary Curator Maurice Broun once called "the one and only predictable thing about hawk migrations," advanced through the area 30 hours earlier.

The front had pushed aside a stagnant, late-summer air mass that had been steaming the area for days. Yesterday, 429 broad-wings were recorded at the lookout. The combined count for the previous two days was a paltry 73 birds. If past is prologue, the big broad-wing push is overdue [5, 6]. If not on the 16th, then certainly a day or two later.

Broad-winged hawks, however, can be as fickle as any bird, and only 2 are sighted during the first 90 minutes of the day's vigil. By 9:00 A.M., things began to pick up. The sky—as if unzipped—is pouring forth with raptors, almost all of them broad-wings, and by 11:00 A.M. more than 500 have been recorded. Most of the migrants are sighted far up-ridge, some as many as 3 to 5 miles off, pepper specks scattered against an overcast sky. Eventually, the specks increase in size, sprout wings, and are identified as broad-wings.

There is more to come, however. Between 11:00 A.M. and noon, 3,302 broad-winged hawks—more broad-wings than breed in all of Rhode Island—are counted swirling above 175 rubber-necking birders at the sanctuary's North Lookout. Three-quarters of a mile away, dozens of latecomers are on their backs in the sanctuary's parking lots, binoculars glued to their eye sockets, trying to catch a glimpse of the flight. Although some of the birds are barely visible more than 1,000 feet above the ridge, many others are passing by at treetop level. Those who have seen the flight in previous years are rejuvenated. Those who have never seen it are awestruck. Dozens of lucky novices have discovered a new calling. Everyone is smiling.

Suddenly, almost as quickly as it had started, the raptor parade dribbles to a standstill shortly after noon. The trailing edge of the massive movement, some 241 broad-wings, is counted between noon and 1:00 P.M. Fifty more pass the North Lookout during the remaining 4 hours of the count. By day's end, 4,291 raptors, representing 10 species of hawks, eagles, and falcons, have been counted at Hawk Mountain's North Lookout. The final tally for broad-wings is 4,118.

The date 16 September 1995 will go down in the sanctuary's record book as the sixth-best one-day broad-wing flight in Hawk Mountain history. Not too surprisingly, the biggest—an astounding 11,349-bird flight—had occurred 47 years to the day earlier, on the 16th of September 1948 (Table 6.2).

The hawk-watchers who visited the sanctuary on 16 September 1995 had every reason to expect a great flight. They were, after all, at the right place at the right time, and they had exactly the right kind of weather. Their ability to predict with reasoned confidence the broad-wing migration at Hawk Mountain Sanctuary that day is one of the great birding detective stories of the twentieth century. To appreciate the story fully, however, one first needs to understand three interlocking aspects of broadwing natural history: breeding phenology, migration geography, and flight strategy. Timing is an essential feature of each.

Plate 5.1 In breeding-plumage, the blackpoll warbler is boldly marked. (Photo by J.Heidecker for Cornell Laboratory of Ornithology)

Plate 5.2 The Allegheny Front Migration Observatory in the mountains of West Virginia. The lines of mist nets in which migrants are captured are in the foreground, and the banding-and-processing building is in the background. The observatory is a popular site for visiting birders. (Photo by Gary Felton)

Plate 5.3 A banded black-poll (note the yellowish feet) has been recaptured. (Photo by C.L. Smith/ Massachusetts Audubon Society)

Plate 5.4 This blackpoll's marathon journey has been interrupted at the Allegheny Front, where it has been weighed, banded, and measured and its age determined. Note the ruffled feathers on the top of its head. The feathers and skin on top of the head have been moistened so that the degree of ossification of the bird's skull can be seen through the skin. Birds hatched during the previous summer still have incompletely ossified skulls during fall migration and can be differentiated thereby from older birds. (Photo by Gary Felton)

Plate 5.5 The Connecticut warbler is one of very few passerines that probably makes the long flight over the western Atlantic from North to South America. (Photo by B.D. Cottrille for Cornell Laboratory of Ornithology)

Plate 6.1 The Kittatinny Ridge as seen from Hawk Mountain Sanctuary's North Lookout in mid-September, at the peak of broad-winged hawk migration in the area. (Photo by K.L. Bildstein)

Plate 6.2 Adult broad-winged hawks are often observed perched quietly beneath the canopy of forest trees. (Photo by B.K. Wheeler/Vireo)

Plate 6.3 Nestling broad-winged hawks near Hawk Mountain Sanctuary in late summer. In less than 2 months these birds will be migrating southward along the Kittatinny Ridge en route to their Central and South American wintering grounds. (Photo by E. Hill/G. Somers Collection, Hawk Mountain Sanctuary)

Plate 6.4 A kettle of broad-winged hawks over Cardel, Veracruz, Mexico, in early October 1995. Approximately one million broad-wings are counted at the site each autumn, with peak passage occurring from 27 September through 6 October each year. (Photo by L.J. Goodrich)

Plate 7.1 Sandhill cranes over the Platte River at sunrise. The Platte River Valley provides unparalleled opportunities for viewing sandhill cranes and, along with the adjacent Rainwater Basin Area, hosts one of the largest concentrations of waterfowl in North America in spring. (Photo by J. Eldridge)

Plate 7.3 (*A*) A wet meadow landscape in the Platte River Valley. It is March and the meadow is occupied by sandhill cranes. (Photo by J. Eldridge)

Plate 7.2 Biologist concealing a cannon net in preparation for trapping sandhill cranes on a wet meadow near the Platte River. Trapped cranes were fitted with radio transmitters so that the biologists could study the birds' habitat use. Cranes were difficult to capture because of their wariness, leading to the use of mounted specimens (such as those shown here) as decoys. (Photo by J. Eldridge)

Plate 7.3 (*B*) A wet meadow landscape in the Platte River. Sandhill cranes meet most of their protein needs by feeding on macroinvertebrates present in wet meadows and grasslands. (Photo by J. Eldridge)

Plate 7.4 Biologists measuring the amount of corn residues still remaining after the departure of sandhill cranes from the Platte River Valley in spring. Sandhill cranes derive nearly all of their energy requirements, and a major part of the fat reserves used in migration and reproduction, from corn gleaned during their spring stay. (Photo by J. Eldridge)

Plate 7.5 The sandhill crane has an extremely varied diet throughout the course of a year. From corn to worms to frogs and baby birds, anything it can get its beak on becomes a meal. (Photo by M. Tremaine for Cornell Laboratory of Ornithology)

Plate 7.6 The Platte River Valley provides vital habitat for many species of midcontinent waterbirds. However, channel shrinkage and loss of wet meadow habitat over the past 60 years have brought the Platte ecosystem to a critical stage. Man's intervention will now be required on a continuing basis to provide appropriate water regimes in order to maintain the existing Platte River ecosystem. (Photo by G. Krapu)

Table 6.2 *The ten best flights of broad-winged hawks at Hawk Mountain Sanctuary, 1934–1995*

Date of flight	Number of birds counted
16 September 1948	11,349
14 September 1978	10,066
18 September 1978	7,222
14 September 1963	6,775
17 September 1968	4,863
16 September 1995	4,118
24 September 1938	4,078
14 September 1958	3,522
17 September 1936	3,398
21 September 1960	3,375

Breeding Phenology

As is true for many long-distance migrants, breeding is hurried in broad-winged hawks, especially for those nesting near the northern limits of the species range in New England and eastern Canada. Broad-wings time their arrival on the breeding grounds to coincide with the springtime snowmelt that uncovers the species' vertebrate prey, an event that usually occurs between late April and mid-May each year. The race to breed successfully in time to beat a hasty retreat to the species' wintering grounds in Central and South America begins shortly thereafter [7, 8].

Male broad-wings are territorial within a day or so of arriving on the breeding grounds, and most individuals court and pair within a week of their appearance in the Northeast (Plate 6.2). Nest construction, which begins immediately thereafter, takes two to four weeks to complete. Females then lay one to four eggs—most clutches consist of two or three—at one- to two-day intervals. Across the center of their Canadian and New England breeding range, most females are incubating eggs by late May or early June. A series of downy nestlings hatches *asynchronously* (not at the same time) 28 to 36 days after incubation begins, usually in late June or early July. Throughout most of their range, nestlings are fed a diet of small mammals, nestling birds, amphibians, and insects [7].

Once hatched, nestlings spend their 1st month in the nest fighting with siblings for food (Plate 6.3). In years when food is limited, some nestlings—usually the younger and smaller ones—starve. Those that do fledge are flying within 5 weeks, typically in late July through early August.

For several weeks thereafter, recently fledged young remain within several hundred yards of the nest, waiting to intercept parents returning with prey. Although some young begin hunting within 6 weeks of hatching, most spend little time in the air—usually less than 15 to 20 minutes a day—during their first two to three weeks out of the nest. Almost all of the young broad-wing's limited flight time is spent within the forest canopy. At 8 to 10 weeks of age—typically in late August to early September—most young of the year are independent of their parents and feeding on their own.

Within several days, these same individuals, together with the entire adult population, will be massing by the thousands as the species begins its annual journey to Central and South America.

The Way South

Most broad-wings begin their southbound migrations alone. Shortly thereafter, however, many coalesce into small groups as the flight converges along a series of more-or-less fixed flyways [9]. Some of the routes are more stable than others.

Flyways along *leading lines*, geographic features that serve to funnel the birds around and through particular landscape features, tend to be the most traditional. Hawk Mountain Sanctuary lies along one such route, a weathered ridge more than 200 miles long and 300 million years old that forms the southeastern border of the Central Appalachian Mountains. Known as the Shawangunk Mountains in New York, and as the Kittatinny Ridge in New Jersey and Pennsylvania, this "endless mountain" of the Lenape Indians, attracts thousands of broad-wings to its slopes each and every fall.

In addition to such north–south trending mountains, large bodies of water also funnel the woodland predators south each autumn. Migrating broadwings are figuratively "hydrophobic" on migration, and few, if any, ever make water crossings greater than 25 miles. Those that do, appear to have been forced to undertake the crossing by local weather conditions.

Most broad-winged hawks nest in the eastern United States and southeastern Canada, and the species' migration is best known for birds breeding east of the 100th meridian. With regard to this population, what appears to happen each fall is shown on Map 6.1.

Hundreds of thousands of broad-wings—the bulk of central Canada's breeding population—skirt the northern and western shores of the Great Lakes, en route to a broad Mississippi River flyway that leads them south to the Gulf Coast of Texas and eventually, into Mexico.

One of the best places to observe this portion of the flight is at Hawk Ridge, Minnesota, a windswept, lakeside dune ridge at the western corner of Lake Superior, just

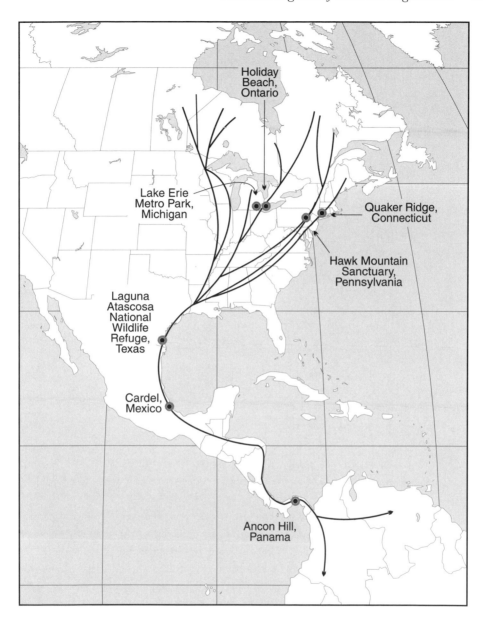

Map 6.1 The southbound migration route of broad-winged hawks in eastern North America. Notable observation points, where large numbers of passing broadwings can be seen, are shown along the route.

outside Duluth. Hawk-watchers at the site, which has been in operation since 1980, have recorded as many as 110,000 broad-wings in a single season. Another major concentration point in the region occurs along the northern shore of Lake Erie, near the mouth of the Detroit River, 20 miles south of Detroit, Michigan. Hawk-watchers at Holiday Beach, Ontario, on the eastern side of the river's mouth, count as many as 110,000 southbound broad-wings each fall; those across the river, at Lake Erie Metro Park, have counted as many as 399,000 in a single season.

Farther east, many of the broad-wings breeding in New England, eastern Quebec, and the Canadian Maritimes travel south to the coastal plains of southern Connecticut. Here, in most years, tens of thousands of birds can be seen at more than a dozen watch sites in the area, including Quaker Ridge, near Greenwich, Connecticut. Many of the birds fly directly over New York City, where counters in Manhattan's Central Park have tallied thousands of broad-wings in recent years.

After the flight crosses the Hudson River, many broad-wings proceed west along a route that eventually positions them over the central Appalachian ridges of northernmost New Jersey and east-central Pennsylvania. There they merge with other individuals that have been following a more inland route to the south. Other broad-wings continue southwest along a more coastal—and decidedly more metropolitan—route which in some years tracks the New Jersey Turnpike to Philadelphia and points beyond.

Throughout eastern North America, the bulk of the season's broad-wing flight passes each established watch site over the course of a few days—usually the same few days—year after year.

Eventually, broad-wings from the eastern and central portions of the species' breeding range converge in massive congregations above the Gulf Coast of Texas, where seasonal tallies at traditional watch sites (stretching from Galveston and Corpus Christi all the way to the Mexican border) approach and exceed 300,000 broad-wings.

Once the migratory stream passes into Mexico, almost every broad-wing in the world continues south in a series of huge flocks (100,000+) along the foothills and coastal slopes of northeastern Mexico's Sierra Madre Oriental (Plate 6.4). By the time the birds reach coastal Veracruz, as many as 1.7 million broad-wings can be seen drifting south above the tiny town of Cardel. A bit farther south, at the 137-mile-wide Isthmus of Tehuantepec, the flight veers to the east along the northern slopes of the Sierra de Tuxtla, en route to the Caribbean coast of Guatemala and the rest of eastern Central America. Although some broad-wings overwinter in Central America, most of the flight squeezes through Middle America's second bottleneck, the 31-mile-wide Isthmus of Panama, where counts from Ancon Hill outside of Panama City range into the hundreds of thousands each year [10].

Once the broad-wings have entered South America, their movements are decidedly less well understood. Anecdotal reports suggest that at least some of the birds proceed south along the western slopes of the Andean Cordillera at least as far as central Colombia, where traditional roost sites include forested slopes of the Combeima Canyon, midway between the Colombian cities of Bogotá and Cali. Although the broad-wings' wintering distribution in South America is not especially well known, they do appear to be common winter residents in portions of forested Colombia, Venezuela, Bolivia, and Amazonian Brazil.

The routes broad-wings take to return to their North American breeding grounds are also considerably less well understood. Most apparently retrace their migratory tracks through Central America and Mexico before fanning across central and eastern North America north of the Rio Grande in late March and early April. Movements across the eastern United States appear to be dispersed relative to those of southbound migrants and, overall, the birds follow a more westerly track in spring than in fall. Tens of thousands of springtime migrants can be expected at traditional watch sites along the southern shores of Lakes Erie and Ontario in late April each year.

Hawk Mountain, by comparison, typically records only a few hundred broad-wings each spring, with most of the flight passing during the third week of April each year.

Flight Behavior

Consider for a minute the challenges facing broad-winged hawks about to embark on fall migration. Successful adults have just completed the burdensome task of raising their young. Young of the year have been on the wing for as little as several hours of their brief lives. The 4,000- to 5,000-mile trip from southern Canada or the northern United States to northern or central South America will take as long as two months to complete, and most of it will be over unfamiliar terrain. The birds need to find energy to fuel their flight. Failure to do so means death.

Migrating birds have several options in regard to fuel for their journey. The first and foremost is to *build up fat*. Because it has a higher caloric content than other potential metabolic fuels and can be stored dry, fat is the metabolic fuel of choice for most migratory birds. Many smaller migrants, including many songbirds and shorebirds, lay down enormous reserves of fat during several weeks of hyperphagia prior to migration each year. Some even double their body mass while doing so (see Chapter 2).

Broad-winged hawks also fatten up prior to migration, but they don't go to extremes. Most add no more than 20% to 40% to their lean body mass; enough to spell a grounded bird during the several weeks of inclement weather it is likely to encounter during the flight, but hardly enough to fuel the entire intercontinental journey. Broad-wings are limited in their ability to accumulate additional fat for migration both by time and by size.

With a lean body mass of just under a pound, broad-wings are relatively large birds; as such, their metabolism is slower than that of smaller migrants. Although this metabolic rate means that it takes broad-wings—gram for gram—longer to burn energy reserves accumulated for migration, it also means that it takes them longer to deposit these reserves in the first place. And as the broad-wing's rushed breeding phenology suggests, many breeders barely have time to raise their young each summer, let alone spend additional time laying down a massive premigratory fat reserve. Even if they had the time to lay down lots of fat, broad-wings would not be able to do so. The problem is one of scale.

To understand this problem, consider what happens when one changes the size of a two-dimensional square drawn on a piece of paper. When the length of each side is doubled, the area inside the square quadruples. When the sides are tripled, the area inside the square increases by a factor of nine. Now, consider what happens when one increases the size of a three-dimensional object such as a bird. Doubling the linear dimensions squares the object's two-dimensional surface area, while cubing its three-dimensional volume.

Thus, if we compare two birds of the same shape and one is twice as big as the other, the larger of the two will have a wing span that is twice that of its smaller companion, a wing area than is four times as large, and a body that is eight times as massive. As a result, even though larger birds have disproportionately larger wing areas than those of similarly shaped smaller birds, the wings of larger birds are decidedly *more heavily wing-loaded* (i.e., carry more weight per unit area) than those of smaller birds. Because of this physical law, lean-bodied, larger birds tend to be closer to their weight limit for effective long-distance flight than smaller birds and, therefore, less able to fatten up in anticipation of it. The broad-wings' relatively large size all but eliminates fat as the single most important source of power for long-distance flight.

A second fueling option available to birds is to *feed and refuel en route*. Passerines, for example, especially those that are normally active by day but that migrate at night, refuel on an almost daily basis by feeding near their daytime roosts. Other migrants, including shorebirds, break up their long-distance movements by spending several days to several weeks refueling at traditional stopover sites en route (see Chapters 8 and 9). Broad-wings have a difficult time employing either strategy.

Broad-wings are quintessential sit-and-wait predators, energy minimalists that meet their metabolic demands not by actively searching for prey, but by spending long periods of time perching quietly, waiting for prey to come to them. Successful perch hunting requires a) an exceptional knowledge of the prey populations within a well-known territory and b) a lot of time; neither is readily available to migrating broad-wings. Even so, these constraints don't mean that broadwings forgo feeding entirely while on migration.

During migration, broad-wings are in the air mainly between 10:00 A.M. and 4:00 P.M., providing them with some time to hunt each morning and afternoon. Because they often are traveling through North America at the same time that many forest passerines are making similar journeys, broad-wings sometimes find themselves surrounded by naive young-of-the-year prey as well as adults out of their normal territories, both of which can make for easy pickings.

During the fall of 1981, Hawk Mountain counters reported that 8% of the more than 600 broad-wings they had seen at close range had obviously distended crops, suggesting that they had fed earlier in the day. An additional 4% were actually seen catching or feeding on flying insects at the time, most of which were migrating dragonflies [11]. My observations suggest that most of the latter predation is serendipitous: the insects involved are flying too close to the broad-wings for the birds to pass them up. In most instances, a soaring or gliding broad-wing simply folds it wings and "sideswipes" its victim, which within minutes has been dewinged and consumed in flight. Feeding events such as these, however, are not likely to provide sufficient energy to fuel the broad-wing's long-distance migration.

How, then, if not with fat and if not by regularly refueling en route, do broadwings power their migratory journeys south each fall? Broad-winged hawks manage to fly between North and South America each year by performing a bit of ecological slight of hand. Rather than depending upon predation and metabolic fuels to power their flights, broad-wings extract the energy needed to do so directly from the atmosphere. These extraordinarily energy-efficient raptors hitch what amounts to a free ride to and from the tropics each fall and spring by soaring most of the way. Indeed, soaring is so important to migrating broad-wings that it is safe to say that nothing about the species' migration ecology makes sense except in the light of this flight behavior.

By definition, soaring occurs when a bird extracts the energy needed for flight directly from the atmosphere. Birds can do so in one of three ways: *Slope soaring* takes advantage of the upward deflection of air caused by mountains and hills. *Thermal soaring* takes advantage of the differential heating of the earth's surface that results in pockets of warm air rising through cooler air. *Dynamic soaring* takes advantage of wind shear and the resulting increase in wind speed with height that occurs over large flat surfaces, such as lakes and oceans. Slope and thermal soaring, which occur mainly over land, require vertical air movements. Dynamic soaring, which occurs mainly over large bodies of water, does not. Because soaring can occur only in moving air, soaring birds (including broad-wings) spend much of their time maneuvering to find this essential aerial habitat [12].

Broad-winged hawks rely on both slope and thermal soaring to achieve their migratory goals. Although they are masters of both, the species is especially dependent upon thermal soaring. Slope soaring serves, more or less, as a backup strategy.

Thermal production depends directly upon the differential heating of the earth's surface that occurs each day as a result of incoming solar radiation. By midmorning on all but the cloudiest of days, the absorption of solar radiation begins to heat the surface of the earth. As it does, air in direct contact with the surface is warmed by conduction. Once sufficiently warmed, this hotter air begins to rise through the colder air above it.

Different land surfaces warm at different rates. Those with *lower surface albedo* (i.e., those that reflect less and absorb more solar radiation—darker surfaces, for example) warm more quickly than those with high surface albedo. Dry surfaces (where evaporative cooling is not possible) warm more quickly than moist surfaces. As a result, landscapes that consist of a patchwork of habitat types, including those that occur throughout most of eastern North America, are likely to produce a series of isolated columns, or thermals, of rising warmer air. Thermals tend to be strongest in late morning through midday—after the sun has risen sufficiently to warm the earth's surface and before strong afternoon winds pull them apart.

In late spring and summer in North America, many thermals rise a mile or so above ground before dissipating. Because they are fueled by sunlight, thermals are far more common on the longer summer days than at other times of the year. My colleagues Paul Kerlinger, Sid Gauthreaux, and Ken Able have used radar to study broad-wings soaring in such updrafts, both in central New York State and in southern Texas. These three biologists recorded broad-wings rising in thermals at rates of 200 to 650 feet per hour; fast enough to allow broad-wings to use this free ride to soar and glide to and from South America each year [13].

Even in summer, however, thermals do not form each and every day. Rainy or cloudy weather, for example, can severely impede the formation of these atmospheric disturbances. Thus, if broad-wings depended solely upon this source of atmospheric energy to fuel their migrations, they might never reach their wintering grounds. For this reason, many broad-wings use slope soaring as a backup. Generally restricted to hilly or to coastal regions, slope soaring allows broadwings to migrate at low cost when thermals are not available.

Slope soaring is possible when sufficiently strong horizontal winds strike an elevated surface—a mountain ridge, for example—and are deflected up and over it. This is exactly what happens along the more-than-200-mile-long Kittatinny Ridge which forms the spine of Hawk Mountain Sanctuary. Each fall tens if not hundreds of thousands of birds use the updrafts associated with this famous ridge, together with the region's thermals, to alternately swirl and surf above the sanctuary's forests—and birders—below.

Broad-wings use several cues to find "good air." Dust and debris are often carried aloft in thermals, and there is every reason to believe that broad-wings, like glider pilots, use such circling masses of reverse detritus to locate vertical air. Because prevailing winds in eastern North America tend to come from the west, mountain ridges that are oriented north–south also are likely to attract broad-wings on migration. The greatest attractant of all, though, appears to be another broad-winged hawk.

More than any other eastern raptor, broad-winged hawks migrate in flocks, the size of which depends upon the magnitude of the day's flight. At Hawk Mountain Sanctuary, for example, "100-broad-wing days" produce flocks of from several birds to several dozen birds, while "1,000-bird days" produce flocks of hundreds of birds. Farther south in Veracruz, Mexico, where a season's broad-wing flight can easily exceed 1 million birds, flock of tens of thousands of the soaring buteos are common.

Broad-wings mass on migration not because the species is more social than other raptors, but rather because doing so enhances the birds' ability to find and use thermals quickly and efficiently. By fanning across the landscape within sight of one another, a group of broad-wings simultaneously samples the atmosphere for pockets of "upwardly mobile" air.

There are no permanent leaders or followers in a flock of migrating broad-winged hawks, simply a series of lucky birds who, having found a column of rising air or a ridge-induced updraft, quickly find themselves surrounded by newfound and decidedly ephemeral friends, eager to take advantage of their "leader's" new resource. In thermals, flocks remain together only until the highest individuals reach the top of the column of rising air, at which point the broad-wings begin to spill out of the vortex and, in hawk-watching parlance, "stream"—tails and wings partially tucked, head-to-tail—to the next available thermal, in which the flock will follow yet another leader to the top.

Groups of spiraling broad-wings are called *kettles*. Although the etymology of the term remains unclear, Hawk Mountain Sanctuary counters have been using it for decades. The term may be derived from the fact that broad-wings lofting in thermals resemble steam rising above a kettle of boiling water. Whatever its origins, calling out an approaching "kettle" of broad-wings at Hawk Mountain's lookout in September electrifies the crowd.

In eastern North America, most of the thermals encountered by migrating broad-wings are relatively narrow columns of rotating air—some measure no more than 10 to 20 yards across—in which soaring broad-wings need to pivot to stay inside. Soaring effectively in narrow thermals requires a tight turning radius. Broad-wings can accomplish this feat because they are very lightly wing-loaded (i.e., relatively big-winged for their body mass) compared to other hawks. The average broad-wing, for example, carries about 355 g (12.5 oz) of body mass for each square foot of wing area; approximately 20% less than a Cooper's hawk, and 30% less than a red-tailed hawk. Being lightly wing-loaded gives the birds more lift, which allows them to fly more slowly in thermals, thereby enabling them to circle more tightly.

Light wing-loading, however, is not without cost. Additional lift makes it easy for broad-wings to get caught up and carried away in thermals. As a result, some broad-wings—especially inexperienced, recently fledged young—can be driven off-course. Most of the broad-wings seen at Cape May Point, the southern peninsular tip of coastal New Jersey, for example, are juvenile birds, individuals that are far off-course and that will need to backtrack along the eastern shore of the Delaware Bay—the body of water that separates the Garden State from Pennsylvania and Delaware—until it narrows sufficiently to permit a short-distance water crossing using powered flight. In most circumstances, experienced adults are far less likely to make this type of costly mistake.

The second trade-off is that birds with low wing-loading glide more slowly on fixed wings than do heavily wing-loaded birds. Gliding between thermals at relatively high speeds is important when one is trying to fly across the better part of two continents. Broad-wings circumvent this obstacle by modifying their flight silhouette when gliding. Rather than assuming the species' characteristic soaring silhouette of fully spread wings and tail, broad-wings partially fold their wings and tuck their tails when streaming out of and into thermals, increasing their wing-loading by 10% to 20%, thereby increasing significantly the speed at which they reach the next thermal. Although flexing their wings increases their rate of drop as well, the trade-off of "lost lift" versus "increased speed" is a successful strategy for the species, provided that thermals are plentiful and interthermal distances are not too great. Which brings me back to the 16th of September.

Why Hurry?

Broad-winged hawks are not the Hawk Mountain Sanctuary's earliest migrating raptors. The midpoint of the bald eagle flight is three days earlier than that of the small buteo. But bald eagles—and, for that matter, all of the other 14 raptor species that migrate past the sanctuary—have far more protracted periods of migration at the site than broad-wings. For example, it takes 16 days for the middle third of the sanctuary's bald eagle flight to pass the North Lookout each fall, and an average of two weeks for the sanctuary's other 14 species to do the same. Broad-wings manage the

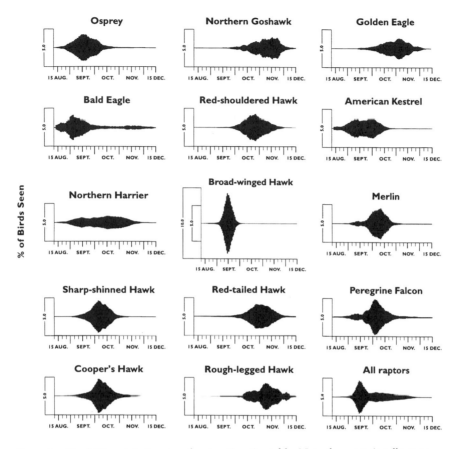

Figure 6.1 Hawk Mountain Sanctuary's migration timetable. Note the exceptionally acute character of the broad-wing flight.

feat in little more than four days each fall (Fig. 6.1). In this respect, at least, broad-wings are exceptionally synchronous migrants.

It is relatively easy to understand why broad-wings don't show up in large numbers at the sanctuary until mid-September each fall. Many broad-wings breeding north and east of Hawk Mountain each year (i.e., those that constitute the sanctuary's annual flight) are still raising young or growing up in mid- to late August. It takes these birds several weeks to acquire their modest fat loads and to complete their journey to central Pennsylvania. Birds that leave earlier than the majority are likely to find themselves searching for thermals on their own, something that may require expensive powered flight and waste precious metabolic fuel.

Why the bulk of the flight passes so quickly, however, is a bit less obvious. In an average year, half the season's flight has passed the sanctuary's North Lookout by the 16th of September. One week later, 95% of the year's broad-wings will have been counted. And by the end of September, more than 99% of the flight will have passed the North Lookout. No other sanctuary raptor is even close to having completed its flight by late September. Indeed, with the exception of broad-wings, only two of Hawk Mountain's migrant species will have concluded even half of their flights by this date. Those two species, the bald eagle and the osprey, have completed 76% and 80% of their flights, respectively, by the end of September.

Some researchers have suggested that broad-wings need to complete their migration through the region earlier than other raptors because a) their preferred prey on the breeding grounds becomes unavailable earlier in the season than the prey of other migrants or b) they need to travel farther. Neither of the proposed scenarios, however, provides a reasonable explanation for the flight's astonishing synchrony each year, resulting in what appears to be a forced migration across all of North America.

Analysis of 60 years of Hawk Mountain Sanctuary flights has led me to conclude that broad-wings rush past the sanctuary, and out of northeastern North America in mid- to late September each year, not because they are running out of food, or because they have so far to travel, but because they are running out of daylight.

Broad-wings migrating past Hawk Mountain on the 16th of September have 12.5 hours of available sunlight. Those passing the sanctuary two weeks later have less than 11 hours and 50 minutes. This, together with the fact that the sun is considerably lower in the sky in late September than it is earlier in the month, and that maximum daily temperatures in the region have dropped by more than 12° F during this period, means that fewer thermals will be available to the birds. And thermals are the fuel tanks of broad-wing migration.

In a very real sense, broad-wings migrating past Hawk Mountain Sanctuary in mid-September each year are racing with the sun. Those that fail to escape the Northeast before thermals become a not-so-predictable resource will need to spend more time in powered flight to reach their final destination. And powered flight is too expensive for long-distance broad-wing migration.

Aeronautic equations developed by the British ornithologist Colin Pennycuick, together with information on the general metabolic demands of broad-wings and their overall levels of activity, make it possible to estimate the amount of energy an individual broad-wing saves by soaring en route to South America each autumn. By my calculations, nonmigratory broad-wings burn about 65 to 75 calories a day, both during the breeding season and on the wintering grounds. Assuming that the same bird flies all the way from southern Canada to central Brazil at its most efficient air speed (i.e., 30–35 mph), it would need to travel two to three hours a day to complete its journey of 4,000 to 5,000 miles in two months. Migration using powered flight almost doubles the bird's metabolic needs each and every day of the flight. A broad-wing using this method needs to find more food in less time, and in unfamiliar territory, than it does on either its breeding or its wintering grounds.

On the other hand, assuming the bird uses soaring rather than powered flight to complete 80% of its migration — which seems reasonable given what is known about the species flight behavior en route — then the bird's metabolic needs increase by only about 20%, which, while not inconsequential, can be met.

Because thermals are so important to migrating broad-wings, it seems reasonable to ask why the species doesn't return to its breeding grounds earlier than it does, when doing so would provide individuals with a bit of a "thermal buffer" in case the weather was bad in early September or mid-September. Doing so, however, would require an earlier arrival on the breeding grounds — which, in itself, creates problems. Advancing spring migration by several weeks means that broad-wings would arrive in North America before the time of dependable springtime thermals. And even if the buteos made that journey, many would arrive on their breeding grounds before snowmelt was complete and small mammal prey was readily available.

Watching Hawk Mountain's Broad-wings

Two things combine to make Hawk Mountain a great place to watch large numbers of migrating broad-winged hawks: the mountain itself and the region's weather.

Hawk Mountain's Kittatinny Ridge all but ensures that at least some broad-wings will pass the sanctuary's North Lookout each fall. This ancient corduroy hill — which stretches from the Hudson River Valley of southern New York almost all the way to the Mason-Dixon line just north of Hagerstown, Maryland — is a perfect highway for updraft- and thermal-seeking broad-wings.

The updrafts produced by winds striking the southeasternmost of the central Appalachian corduroy hills, together with local thermals, allow many of the broad-wings that are sighted at the Hawk Mountain Sanctuary to fly for miles at speeds of up to 35 mph without beating a wing. Indeed, it is the juxtaposition of these two sources of free energy — slope and thermal soaring — that all but guarantees that at least a few thousand broad-wings will be counted at the sanctuary's North Lookout each autumn. Just how many will pass each year, and when exactly the flight will occur, depend upon the weather — not only at the sanctuary, but elsewhere as well.

As the region's southernmost ridge of consequence, the Kittatinny is the last chance southbound broad-wings have to fall back on slope soaring during this portion of their southbound migration. Slope soaring, however, is possible only when there are winds; at Hawk Mountain Sanctuary, the best winds come out of the northwest.

The association between northwest winds and good flights at Hawk Mountain Sanctuary has been known for a long time. After a single year at the site, Maurice Broun, the sanctuary's curator, was "guaranteeing good flights" to visitors the day after a cold front had passed. And the passage of a cold front is almost always followed by northwest winds. Several years ago, Paul Allen, Laurie Goodrich, and I decided to quantify Broun's guarantee [6].

The three of us restricted our analysis of the association to the period between 23 August and 30 September each year, the time when 98% of the broad-wings typically pass the sanctuary's North Lookout. We used two data sets in our analyses: the sanctuary's daily counts of broad-winged hawks from 1934 through 1991, and the U.S. Weather Bureau's accounts of cold fronts in the region during that time. Our initial analysis revealed that cold fronts passed the sanctuary once every four and a half days, and that broad-wings were counted at a rate of 283 birds per eight-hour day of observation. Once we combined the two data sets, we discovered that almost twice as many broad-wings—327 birds per day versus 182—were counted during the three days following the passage of a cold front than at any other period. Broun was right: cold fronts do bring good flights.

Over the years Hawk Mountain has had its share of good flights. Indeed, a thousand or more broad-wings have been sighted on a single day 134 times at the sanctuary. And in almost every instance, the flight had been preceded by a cold front. Although thousand-bird broad-wing days have occurred at least once for each and every date from the 8th to the 28th of September, more than half of Hawk Mountain's red-letter broad-wing days have occurred between the 15th and 20th of the month. And with an average passage rate of 95 broad-wings per hour, the 16th of September stands at the top of the list: 46,917 broad-wings—almost 10% of all that have ever been seen at the sanctuary—have been counted on that date alone.

Tropical storms rarely reach the Appalachian Mountains of eastern Pennsylvania. When they do, they can have a dramatic effect on the year's flight of broad-winged hawks. Only once in the sanctuary's 60-year history have more than a thousand broad-wings passed the North Lookout on each of four consecutive days. The single run occurred between 23 and 26 September 1938, shortly after the Great Hurricane of 1938 devastated much of coastal New England. Maurice Broun credited the storm, which had passed several hundred miles east of the sanctuary two days earlier, with the exceptional flight. A somewhat similar six-day flight of more than 6,500 broad-wings, occurred several days after the passage of Tropical Storm David in 1979. Both storms, which passed east of the sanctuary, almost certainly sent many broad-wings, which would otherwise have passed south and east of the sanctuary, up and over the central Appalachians and onto the Kittatinny Ridge flyway.

Hurricane Fran, on the other hand, which passed west of Hawk Mountain in early September 1996, all but eliminated the sanctuary's broad-wing flight that year. The resulting seasonal total of 1,809 broad-wings—which shattered by more than a thousand birds the previous seasonal low of 2,886 broad-wings set in 1946— amounted to fewer broad-wings than have been reported on 60 single-day counts at the sanctuary. That this cyclonic storm, which had begun as an unpretentious tropical depression off the coast of West Africa several weeks earlier, could so completely dominate the migratory geography of an endemic North American raptor helps demonstrate just how small a planet Earth really is.

The Surprisingly Brief Ornithological History of the Flight

Given what is now known about the fall migration of the broad-winged hawk, it may come as a surprise that this significant avian event eluded ornithologists for most of the nineteenth century. In retrospect, the delayed discovery of the flight is not surprising.

The broad-wing's secretive nature during the breeding season, together with the fact that broad-wings tend to occur at low densities, even in the best of habitats, made it difficult for even experienced nineteenth-century ornithologists to find and study this raptor. As a result, relatively little was known about any aspect of the species' ecology until after the turn of the century.

Alexander Wilson, for example, devoted most of his early nineteenth-century account of the bird to the broad-wing's plumage and associated feather lice. John James Audubon focused a good deal of his text on the idiosyncratic behavior of an orphaned nestling and gunshot adult he had secured as subjects for his paintings. And while both Wilson and Audubon noted the species' habit of circling in flight— now a well-known aspect of the broad-wing's migratory flight—neither of these ornithologists, nor any of their contemporaries, ever hinted at the bird's spectacular migration behavior.

In part, information on the broad-wing's migratory habits was lacking because the birds were difficult to locate during the breeding season, even in areas where they were quite common, making a lack of sightings in winter unsurprising. In part, it was lacking because the species tended to migrate along largely inland corridors, far from the large bodies of water that were then most frequently visited by ornithologists. In part, it was lacking because the flight was concentrated over a few easily missed days each fall and often occurred at heights that would make the flight difficult to see with unaided eyes.

Thus, many local and regional authorities of the day concluded that broad-wings were yearround residents in the ornithologists' geographic areas of expertise. Even Witmer Stone, who in 1937 described the broad-winged hawk as a major participant in the Cape May's "great flights of autumn" [14], had described the species as a non-migratory "rare resident" in 1894 [15].

In fact, it was not until almost 1880 that the ornithological community "discovered" the broad-wing's migratory behavior. The breakthrough came long after the far less spectacular migratory movements of many other New World raptors had been described in some detail. In retrospect, the delay was understandable. The use of field glasses—and then binoculars—certainly helped, as did the advent of the field guide.

The earliest account of broad-wing migration I have been able to uncover is that of the oil-rig salesman and amateur ornithologist George B. Sennett, published in 1879. Sennett, who at the time was considered "the" authority on the birds of the Rio Grande Valley, described the movements of at least 50 broad-wings on spring migration in Gulf Coast Texas as being "easy, graceful, and at times, quite rapid," quite the opposite of the bird's heavy and sluggish subcanopy flight during the breeding season [16]. This was a different bird than the one most ornithologists who had studied it during the breeding season were familiar with, and George Sennett was one of the first to recognize its migration-period transformation in behavior.

By the mid-1880s, Charles C. Trowbridge was reporting "immense clusters" of hundreds of migrating broad-wings in and around New Haven, Connecticut, including a "great flight" on 16 September 1887 that began shortly after 9:00 A.M., 108 years earlier to the hour than the Hawk Mountain Sanctuary flight described at the beginning of this chapter [17]. Over the next few years, other field-glass-toting ornithologists joined the search, and by century's end the general nature of the species' migratory movements north of the Rio Grande—its movements near Detroit along the northern shores of Lake Erie, and near Rochester, New York, along the southern shores of Lake Ontario, as well as over Montclair, New Jersey, east of New York City, and along the Kittatinny Ridge in the vicinity of the Delaware Water Gap—was well known to ornithologists and bird-watchers of the day. At about the same time, the broad-wings also became known to the region's gunners, an awareness that eventually led to the ornithological discovery of the flight at Hawk Mountain.

Although hawk-shooting apparently was widespread along much of the Kittatinny Ridge in the late 1920s and early 1930s, reports from that time indicate that the most popular location for this mainly Sunday sport was the place the gunners called Hawk Mountain, an easily approached escarpment along a mountaintop road connecting Eckville and Drehersville, Pennsylvania (Fig. 6.2).

Many of the shooters were anthracite coal miners, who visited the ridge each Sunday after having worked in the mines for six days. Although the Pennsylvania Game Commission prohibited game-hunting on the Sabbath, there was no such restriction on "vermin." And raptors—especially those suspected of taking ruffed grouse, the state's premiere upland game bird—were then considered vermin. A $5 Game Commission bounty on the northern goshawk made that species an especially lucrative target. Estimates from the era suggest that hundreds of raptors, including eagles, ospreys, falcons, accipiters, and harriers, as well as broad-wings and other buteos, were being shot on single weekends. By the early 1930s, shooting at the site was so extensive that brass from discharged cartridges was being collected and sold for scrap metal.

All of this changed in June of 1934, when the conservationist Rosalie Edge, having seen photographs of the events that previous fall, traveled to the site and purchased an option on the property (Fig. 6.3). That August, Edge hired the Vermont naturalist Maurice Broun as "ornithologist-in-charge" of the newly established refuge for birds of prey, the first of its kind anywhere. Broun spent part of his first

Figure 6.2 Prior to the Hawk Mountain Sanctuary's founding in 1934, shooting hawks was a common weekend practice at Hawk Mountain during fall migration. (Photo by R.H. Pough)

Figure 6.3 Hawk Moun-
tain Sanctuary founder
Rosalie Edge. (Hawk
Mountain Sanctuary)

fall at the site posting the newly named Hawk Mountain Sanctuary and confronting
local shooters. He also began recording the migration.

There was much publicity surrounding the sanctuary's formation, and that fall
500 bird-watchers and naturalists flocked to the "newly discovered" wildlife refuge.
By 1950, Hawk Mountain was attracting more than 10,000 visitors annually. By the
mid-1990s, the number of visitors had increased to 80,000. The sanctuary, which was
once known mainly for the numbers of birds counted at its lookouts, is now equally
famous for the numbers of people that visit each fall.

Today, Hawk Mountain Sanctuary—110 miles west of New York City and 85 miles
northwest of Philadelphia—is a member-supported, not-for-profit international
center for conservation, education, and research. Throughout its history, Hawk
Mountain has been a leader in science-based raptor conservation. The sanctuary cur-
rently maintains the longest and most complete record of raptor migration in the

world. This extensive database played a key role in exposing first-generation organochlorine pesticides, including DDT, as causative agents in the precipitous decline in populations of bald eagles, peregrine falcons, and other species of birds of prey that occurred earlier in the twentieth century, as well as in recording subsequent rebounds in these populations following decreases in the use of the pesticides.

Recognizing the need to protect raptors throughout their migratory journeys, in 1988 the sanctuary launched "Hawks Aloft Worldwide," a cooperative global-conservation initiative that uses the plight of threatened raptors, together with the spectacle of their migrations, to engender local support for raptor conservation worldwide. One of the project's earliest successes was with the broad-winged hawk. Although the species is no longer hunted in Canada and the United States, it is not so lucky elsewhere in its range. This is particularly so in the western cordillera of the Colombian Andes, where local lore suggest that killing broad-wings hastens the passage of Lent and that fat rendered from their carcasses has medicinal value. As a result, as recently as the early 1990s, large numbers of North American–bound broad-wings were being killed each spring by local campesinos who were shooting the birds at traditional nocturnal roost sites. Today, thanks to a major education campaign led by local Hawks Aloft Worldwide cooperators from Bogotá and Tolima, the annual slaughter has all but been eliminated, and local residents now celebrate the birds' passage each April with a festival.

References

1. Broun, M. 1949. Hawks Aloft: The Story of Hawk Mountain. Cornwall (NY): Dodd, Mead.
   A captivating account of the Hawk Mountain Sanctuary's early years as told by its curator.
2. Harwood, M. 1973. The View from Hawk Mountain. New York: Scribner's.
   A hawk-watcher's account of the sanctuary's activities in the 1960s and early 1970s.
3. Brett, J.J. 1991. The Mountain and the Migration. Ithaca (NY): Cornell Univ. Pr.
   A practical guide to Hawk Mountain Sanctuary's history, geology, wildlife, and migrating raptors.
4. Bednarz, J.C., D. Klem, L.J. Goodrich, and S.E. Senner. 1990. Migration counts at Hawk Mountain, Pennsylvania, as indicators of population trends, 1934–1986. Auk 107:96–109.
   A technical account of the numbers of raptors seen at the sanctuary between 1934 and 1986, especially in relation to organochlorine pesticide use in North America.
5. Allen, P.E., L.J. Goodrich, and K.L. Bildstein. 1995. Hawk Mountain's million-bird database. Birding 27:24–32.
   The history of the sanctuary's database together with a temporal field guide to raptor migration at the site.
6. Allen, P.E., L.J. Goodrich, and K.L. Bildstein. 1996. Within- and among-year effects of cold fronts on migrating raptors at Hawk Mountain, Pennsylvania, 1934–1991. Auk 113:329–338.

A detailed examination of the relationship between weather and raptor migration at Hawk Mountain Sanctuary.

7. Goodrich, L.J., S.C. Crocoll, and S.E. Senner. 1996. Broad-winged hawk (*Buteo platypterus*). In The Birds of North America, No. 218 (A. Poole and F. Gill, eds.). Washington, DC, and Philadelphia: Am. Ornith. Union and Acad. Nat. Sci.

The most up-to-date life history of the species.

8. Dunne, P. 1995. The Wind Masters: The Lives of North American Birds of Prey. Boston: Houghton Mifflin.

A series of engaging "bird's eye" views of North American raptors, including a chapter depicting the thoughts of a migrating broad-winged hawk.

9. Heintzelman, D.S. 1979. A Guide to Hawk Watching in North America. University Park (PA): Keystone.

A useful introduction to hawk identification and to the locations of important hawk-migration watch sites in North America.

10. Smith, N.G. 1980. Hawk and vulture migration in the Neotropics. In Migrant Birds in the Neotropics: Ecology, Behavior, Distribution, and Conservation (A. Keast and E.S. Morton, eds.). Washington, DC: Smithsonian Inst. Pr.

11. Shelley, E., and S. Benz. 1985. Observations of aerial hunting, food carrying, and crop size in migrating raptors. *In* Conservation Studies on Migrating Raptors (I. Newton and R.D. Chancellor, eds.). ICBP Tech. Publ. 5. Cambridge (Eng): ICBP [International Council for Bird Preservation].

A brief account of raptors, including broad-wings, feeding during migration at Hawk Mountain Sanctuary.

12. Kerlinger, P. 1989. Flight Strategies of Migrating Hawks. Chicago: Univ. Chicago Pr.

A technical but accessible description of the mechanics of raptor migration.

13. Kerlinger, P. 1985. Seasonal timing, geographic distribution, and flight behavior of broad-winged hawks during spring migration in South Texas: a radar and visual study. Auk 102: 735–743.

An informative account of the movements of broad-winged hawks through the coastal plains of southeastern Texas.

14. Stone, W. 1937. Bird Studies at Old Cape May. Philadelphia: Delaware Valley Ornithol. Club.

15. Stone, W. 1894. The Birds of Eastern Pennsylvania and New Jersey. Philadelphia: Delaware Valley Ornithol. Club.

Interesting historical accounts of bird distribution and migration in the Cape May area.

16. Sennett, G.B. 1879. Further Notes on the Ornithology of the Lower Rio Grande of Texas, from Observations Made during the Spring of 1878. Bull. U.S. Geol. Surv. Territories, Washington, D.C. Volume 5.

A typical account of early natural history observations from the American frontier.

17. Trowbridge, C.C. 1895. Hawk flights in Connecticut. Auk 12:262–265.

An early descriptive account.

7 GARY L. KRAPU

Sandhill Cranes and the Platte River

High horns, low horns, silence, and finally a pandemonium of trumpets, rattles, croaks, and cries that almost shakes the bog with its nearness, but without yet disclosing whence it comes. At last a glint of sun reveals the approach of a great echelon of birds. On motionless wing they emerge from the lifting mists, sweep a final arc of sky, and settle in clangorous descending spirals to their feeding grounds. A new day has begun on the crane marsh. . . . Our ability to perceive quality in nature begins, as in art, with the pretty. It expands through successive stages of the beautiful to values as yet uncaptured by language. The quality of cranes lies, I think, in this higher gamut, as yet beyond the reach of words.
—Aldo Leopold, *A Sand County Almanac*, 1949

What could one add that might better capture the wildness and magnificence of sandhill cranes? If you have not seen the great spring concentration of cranes along the Platte River in Nebraska, you should do so. It is one of the world's great spectacles of animal migration and it provides an indelible experience on many levels. The visual and aural immensity of it saturates the senses. At the same time, it is interesting and important to understand the dynamics of the phenomenon, the driving but often subtle interactions between human activities and those of the cranes. The detailed, long-term studies by Gary Krapu and his colleagues reveal that the cranes are not only dancing with one another; they are enjoined by fate to engage in an extended pas de deux with us, and that dance has just begun. Whether they will still mass on the Platte a hundred or a thousand years from now is largely in our hands. —*K.P.A.*

There are few places as captivating in early spring to a student of bird migration as the Platte River Valley of south-central Nebraska between Grand Island and Kearney. Here, at dawn from mid- to late March, one can view thousands of sandhill cranes departing their roosts located in channels of the Platte River (Plate 7.1) or in nearby fields during the rest of the day. Moreover, the cranes are often close to roads, offering one of the best opportunities on the continent for viewing this usually wary bird. For me, spending time among the cranes in spring is special because the species has lived in what is now Nebraska for more than 1 million years. Listening to sandhill cranes and observing their behavior, one is left contemplating conditions in the Pliocene that would have fostered the evolution of such a bird.

My first visit to the Platte was in the early spring of 1978. I had agreed to serve as project leader for the Platte River Ecology Study which the U.S. Fish and Wildlife Service had undertaken to gain insight into the habitat requirements of migratory bird populations using the Platte River Valley, particularly sandhill cranes [1, 2]. Intrigued by the species' migratory habits, I wanted to know a) why nearly 80% of all sandhill cranes in North America gather along the Platte and North Platte rivers in spring and b) what were the cranes' habitat needs during their stay. In this chapter, I draw upon information gained from my research and from that of my colleagues to identify underlying factors responsible for the current springtime staging pattern of sandhill cranes in the Platte River Valley. I conclude with a description of efforts under way in the Platte River Valley to protect the habitat that supports sandhill cranes and other waterbirds and, in that context, I discuss our information needs.

The Great Plains, a Landscape in Transition

Until the late 19th century, the area known as the Central Plains was a vast prairie wilderness. The fertile Platte River Valley with its subirrigated meadows and wide channels attracted great herds of bison and an abundance of migratory waterbirds in spring. The wide channels of the Platte were a favored roosting area for sandhill cranes during their northward migration. Before agriculture and crop residues, food availability was variable from year to year and cranes were more opportunistic in their migratory habits. Thus, flocks were more dispersed than in modern times, and stopovers were probably of shorter duration.

Up to 1850, The Great plains remained sparsely inhabited, mostly by Indian tribes which represented the last in a long line of hunters and gatherers that had occupied the plains during the postglacial period (10,000–12,000 years and possibly longer). At low population densities and carrying primitive weapons, humans inhabiting the region before the mid-19th century likely had little effect on sandhill crane populations that nested in or migrated across the central and northern Great Plains.

Cataclysmic change in the prairie ecosystem began in the third quarter of the 19th century. Treaties that had left large parts of the Great Plains in a natural state under the control of indigenous tribes were broken as political pressure mounted to open the region to white settlers, miners, and others. Confiscation of Indian lands was backed by military force in the 1860s and 1870s, ending the nomadic life of the Plains tribes. This action set the stage for the arrival of millions of homesteaders and for agricultural development of the Great Plains that would profoundly affect the distribution and habits of sandhill cranes in the coming century.

Sandhill cranes, because of their large size, were particularly vulnerable to hunting during settlement of the Great Plains between 1880 and 1910. Although habitat remained plentiful for cranes, without laws to control hunting and with a lack of sanctuaries the Plains breeding population was decimated by the turn of the century. Northern-nesting populations fared better but their numbers also declined.

Concerns raised by the decline of many species of waterbirds owing to uncontrolled hunting led to the introduction in Congress in 1904 of legislation to place migratory birds under federal protection. After nearly a decade of intense lobbying by conservation leaders, the Migratory Bird Act was passed by Congress and signed into law by President Taft on 4 March 1913. In 1916, the Migratory Bird Treaty between the United States and Canada was ratified for Canada by Great Britain, expanding protection for sandhill cranes, waterfowl, and many other species to a large part of North America. The U.S. Biological Survey was given responsibility for enforcing the treaty.

With the gradual decline of uncontrolled hunting, the midcontinent sandhill crane population began to rebound. The recovery was aided by the severe drought and Great Depression of the 1930s which reduced human densities markedly across large areas of the Great Plains. National wildlife refuges (NWRs) were being established in most states on the Central Flyway, providing additional protection for sandhill cranes [3].

The Introduction of the Mechanical Cornpicker

Agriculture had provided some food for sandhill cranes from the onset of farming in the late 19th century, but handpicking of corn and threshing of other crops left little grain in the fields after harvest. By the early 1940s, though, mechanical corn-pickers were replacing human laborers for harvesting corn in the Great Plains and leaving from 6% to 8% of the crop in the fields, creating an abundant food supply for cranes and other wildlife. As corn residues increased in the North Platte and Platte valleys close to favored channel-roosting habitat, the number of migrant flocks of sandhill cranes stopping increased as did the length of their stay. In the spring of 1943, an estimated 100,000 sandhill cranes stopped in the North Platte River Valley. Farmers made no effort to chase the growing numbers of cranes from their fields or disturb them on their roosts, for the birds posed no threat to their farming operations.

Corn availability to cranes grew from the 1940s through the 1970s as yields increased. By 1950, most of the corn in the Platte River Valley was being harvested by mechanical cornpicker. From 1956 to 1979, corn yield and availability to cranes and other wildlife in Nebraska increased fivefold, mostly because of the growing use of irrigation, a greater input of fertilizer, and the development of hybrids capable of producing high yields. By the late 1970s, most of the cropland in the valley was in corn, nearly all the corn was irrigated, and corn residues were abundant.

Increasing Corn but Shrinking Channels

The growing abundance of corn residues in the Platte River Valley as a result of the mechanization of agriculture was coupled with a dramatic decline of roosting habitat in the channels of the upper and central Platte River and elsewhere along their migration route, starting in the 1940s [4]. Historically, the Platte River ecosystem was maintained primarily by runoff from the western plains and eastern slope of the Rocky Mountains which was carried through the North Platte and South Platte rivers. By late summer, early accounts indicated, flows in the upper and central Platte River were but a small fraction of spring flows. The combination of spring flooding and summer drying of much of the channel kept sandbars unstable and mostly free of vegetation, thus helping to maintain a wide, shallow, and braided channel. Although trees were already present on some islands in the river channel in the early 19th century, explorers' reports indicated a general lack of trees on the banks. In 1813, Robert Stuart, who was exploring the North Platte River, noted in his diary that, except for a lone tree, no timber existed on the north side of the river for 200 miles upstream from its confluence with the South Platte River [5]. In 1865,

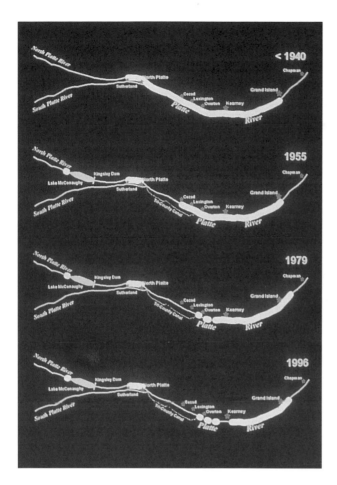

Map 7.1 The changing distribution (*shaded area*) of spring-staging sandhill cranes in the North Platte and Platte river valleys over the past half-century. Because of a progressive narrowing of channels from west to east, starting at the confluence of the North Platte and South Platte rivers, cranes have abandoned much of the channel between Overton and Kearney and all of the Platte River upstream of where return flows from Tri-County Canal discharge back into the river near Overton. (Photo by G. Krapu)

Union Pacific Railroad surveyors reported that the channel width of the Platte River varied from about 770 to 1,870 yards between the confluence of the North and South Platte rivers and where Grand Island is located today [4]. As late as 1938, the channel remained wide through most of the upper and central Platte. At Overton, for example, the channel width was 1,672 yards (1,520 m), only 99 yards (90 m) less than in 1865.

Channel width began shrink precipitously after Kingsley Dam was completed in 1941 on the North Platte River in western Nebraska, creating Lake McConaughy. For purposes of irrigation and power generation, most of the Platte River flows from just below the confluence of the North Platte and South Platte Rivers to near Overton were rerouted into the Tri-County Canal (Map 7.1). As a result, most of the

Figure 7.1 The former width of the Platte River channel of Kearney, Nebraska, is indicated by this corridor of woodland. Channel width in nearly all of the upper Platte River and in large sections of the central Platte has been reduced by at least 90% and can no longer support cranes. (Photo by J. Eldridge)

active channel bed for 61 miles downstream to near Lexington became woodland over the next 2 decades. By 1980, a milewide swath of maturing riparian woodland marked the route of what had been the channel of the upper Platte River (Fig. 7.1). The river channel remained wider downstream from Overton where return flows from the Tri-County Canal reentered the river, but the channel width was still less than 20% of what it had been in 1938.

As river channels narrowed, sandhill cranes gradually abandoned all of the upper Platte River Valley above where return flows from the canal reentered the river. Lawrence Walkinshaw, the first biologist to study sandhill cranes extensively, saw only 7 cranes between Cozad and North Platte in 1953 and none in 1954 when he visited the Platte and North Platte rivers [6]. Cranes still occupied the section between Cozad and Lexington in the late 1950s, but by the 1970s the entire reach from the confluence downstream to Lexington had been abandoned.

With loss of the upper Platte, cranes became separated into 3 staging groups: one is located on the 71-mile section of the central Platte between Overton and Chapman, a short distance east of Grand Island; another is situated upstream in a 30-mile section of the North Platte River to the west of the city of North Platte; and a third is at the upper end of Lake McConaughy (see Map 7.1). Sandhill cranes staging in the Platte Valley are of 3 subspecies, the largest being *Grus canadensis tabida* (greater), followed by *G. c. rowani* (lesser), and by *G. c. canadensis*, the smallest. The breeding grounds of these races remain poorly defined, but the midcontinent population of *G. c. tabida* is thought to breed in northern Minnesota, southwestern Ontario, and adjoining areas in Manitoba. *G. c. rowani* presumably breed across a wide area of central Canada [3], but racial composition across much of the region remains poorly defined. Lesser sandhill cranes breed at higher latitudes in the central and western arctic, but little is known concerning the southern limits of their breeding range.

Seeking Answers

Loss of the upper Platte River Valley as a springtime staging area for the midcontinent sandhill crane population and declining channel width downstream from Overton to Grand Island had, by the 1970s, raised widespread concern that cranes and other migratory waterbirds were about to be displaced from the Platte River, with unknown consequences to these populations. This concern, and continuing expansion of irrigation development, resulted in a growing national debate over water policy issues involving the Platte River Basin. Responding to issues that had been raised, the U.S. Department of the Interior requested and received funds from Congress in 1977 to begin a major study to assess the water needs of migratory waterbirds and man in the Platte River Basin, culminating in the Platte River Ecology Study.

Learning How to Capture Cranes

As a starting point for determining habitat needs of sandhill cranes during the staging period, I decided to radio-mark and systematically monitor individual cranes during their spring stay. Fieldwork began in late February 1978, shortly after the cranes started to arrive. Recognizing that capturing cranes would be difficult in late winter, I sought the expertise of Charlie Shaiffer, a biologist experienced in live-trapping waterbirds. We also drew insight from the work of Federal Game Management Agent Robert Wheeler, who had successfully trapped sandhill cranes in the Platte Valley a decade earlier. Our strategy was to lure the cranes to decoys in open fields where well-camouflaged cannon nets would, at the detonation of powder charges, be propelled over the unsuspecting cranes before they could take flight.

Despite numerous precautions, we were not prepared for the high level of wariness shown by the early-arriving flocks, and our first trapping attempts ended in total failure. Flocks entering fields where our wooden decoys had been carefully positioned not only shied away, but in some cases flew low and carefully inspected the wire that led back to the observation blind used for setting off the charges. With our research program in jeopardy, we decided to move on to plan B: the use of taxidermy mounts as decoys. The only problem was that we had no mounts, prompting a widespread search for mounted cranes. As no one was about to lend us their prized specimens when they learned of our intent, the mounts we were given were a bit bedraggled (Plate 7.2). To our great relief, however, cranes seeing their own kind, notwithstanding a somewhat disheveled appearance, lowered their guard, and trapping operations were able to begin in earnest.

Habitat Preferences

Once cranes had been radio-marked, trackers working from antenna-equipped vehicles began systematically driving the roads of the study area, stopping at regular intervals to listen for the frequencies of transmittered cranes. (The frequencies had been programmed into the memory of the trackers' multichannel receivers.) Trackers attempted to locate each crane once per hour, from early morning before the cranes left their roosts until dusk when the birds returned. Each location was established by triangulation of bearings from the nearest roads. When our monitoring of radio-marked cranes had ended and analyses were completed, we were able to determine that the birds were dividing their time mostly among cornfields (60%), native grassland (28%), and alfalfa hayland (9%). For reasons unknown, cranes were using native grassland and planted hayland more than we had expected.

To gain insight into this higher than expected use of grassland and hayland, we systematically monitored the cranes' activity by habitat type, using standard time-budget techniques. We found that the amount of time cranes spent searching for food varied by habitat, increasing from 20% in cornfields to 30% in native grassland to 44% in alfalfa hayland. From this pattern, we concluded that the foods cranes were seeking in native grassland and haylands required a greater foraging effort, leading to a disproportionate amount of time being spent in those habitats. During migration, sandhill cranes roost in a variety of sites including braided river channels, wet meadows, shallow lakes, and uplands. In the Platte River Valley, early-arriving cranes roost on ice in the river channel or on the frozen meadows when required, but they shift to shallow waters of the channel when possible. Cranes prefer roosting on submerged sandbars in channels greater than 165 yards wide (150 m) and avoid channels less than 55 yards wide (50 m) [7]. Vegetation height on riverbanks and on islands also influences roosting distribution when the channel width is less than 165 yards

(150 m). The preference of cranes for wide channels probably is associated with greater security from predators and other forms of disturbance, security that presumably is particularly important where flocks stage for prolonged periods. Human activity near the river channel also affects crane use. Cranes avoid sections of river that are near bridges and roads. Less than half as many cranes roost in river segments where there are bridges, adjacent roads, or both than in segments without these features. Distribution of wet meadows also influences the staging distribution of sandhill cranes in the North Platte and Platte river valleys. Humans, then, through a variety of actions affecting the landscape of the Platte River Valley and the Great Plains, have strongly influenced the current distribution of cranes and have made the birds' continued presence contingent on active habitat management.

Feeding Ecology

Corn, to no one's surprise, dominated the crane diet in cornfields, but the finding that cranes were feeding almost exclusively on macroinvertebrates in native grassland (including wet meadows) was unexpected (Plate 7.3). Earthworms were the principal animal food taken in both grassland and alfalfa hayland, followed by insects. Snails accounted for nearly 25% of the diet in grassland, but less than 5% in hayland. We concluded that cranes had to forage on invertebrates in native meadows and hayland to obtain protein and calcium; corn, while an excellent source of energy because of its high carbohydrate content, is deficient in those nutrients.

By the 1970s, corn residues were sufficiently abundant to support the needs of nearly the entire midcontinent sandhill crane population as well as large numbers of waterfowl during early spring. Ken Reinecke, a member of the Platte River Ecology Study team, measured corn availability in the fields (Plate 7.4) and found that, on average, about 8% of the standing crop of corn (of which about 25% was shelled kernels) remained in the fields after harvest. Although cattle were turned into many of the fields for fattening during the fall and winter, most of the shelled kernels remained available for waterfowl and cranes. We estimated that a population of about 500,000 cranes would remove about 1,471 metric tons of corn each spring to meet its needs. That amount represented only about 23% of the corn available in fields when the cranes arrived in early spring. As cropland area and corn acreage increased in the 20th century throughout the Platte River Valley, areas of native grassland and wet meadows diminished and macroinvertebrates declined, resulting in a marked disparity between the availability of high energy and corn and protein-rich macroinvertebrates. In the 1970s, cranes we monitored spent as much time each day obtaining the estimated 3% of their diet formed by macroinvertebrates as they did obtaining the 97% of dietary corn. Conspicuous blackened patches on meadows and haylands reflected the intense foraging effort being put forth by cranes probing into the soils in search of soil invertebrates.

Refueling for Migration and Reproduction

Sandhill cranes are lean on arrival in the Platte River Valley in late February and early March. In the late 1970s, we found that lesser sandhill cranes weighing about 3,000 g (6.6 lbs) on arrival acquired about 500 g (1.1 lbs) of fat during their spring stay. After leaving the Platte in early April, the birds stored still more fat during stopovers at the northern edge of the Great Plains in prairie Canada. By the time lesser sandhill cranes arrived on the breeding grounds in the Yukon-Kuskokwim Delta in western Alaska, we found, fat reserves equal to fat acquired in the Platte River Valley were still present. Thus, the pound of fat acquired in the Platte River Valley provided a major nutrient source for reproduction. Juveniles acquired less fat than adults, despite high corn availability, possibly because juveniles spend more time searching for macroinvertebrates in grasslands and haylands and, as nonbreeders, probably require less fat.

Implications of a Changing Landscape

Although corn converted to fat has both fueled the migration and provided a major part of the reproductive energy requirements of midcontinent sandhill cranes throughout the second half of this century, there is no guarantee that corn residues will remain adequate to meet these needs in the future. In fact, changes in staging behavior and other evidence suggest that corn residues and fat storage rates are in decline. In the late 1970s, cranes met their energy needs mostly within 2 or 3 miles of the Platte River; by the 1990s, cranes were ranging several miles beyond the valley to feed, an uncommon sight 2 decades ago. Thus, when a blizzard killed 2,000 cranes in the Platte River Valley in March 1996, I measured the body mass of the storm-killed sandhill cranes and found that body mass in the Canadian race had fallen substantially from similar measurements I had taken during 1978 and 1979. Body mass also averaged lower in the lesser sandhill cranes, but the differences were not statistically significant.

Because fat levels of storm-killed cranes may not have been representative of the population, I initiated new studies in the Platte River Valley in spring 1998 to confirm or refute the hypothesis that fat storage rates had declined over the past 2 decades and, if so, why. Preliminary results from these studies confirm that the fat storage rate may have declined as much as 50% since the 1970s. Although the underlying causes for reduced fat storage are still not known, lower corn residues are probably the main reason for the declining rate of fat storage.

More efficient corn-harvesting methods than those used in the past, growing competition from springtime-staging midcontinent Arctic-nesting geese, and an increase in the number of sandhill cranes present in spring are the most likely factors responsible for the declining fat storage rate among cranes. From the various data sets we are acquiring under current studies, a mathematical model will be developed to predict the number of sandhill cranes that can be supported in the Platte River Valley under current management practices and alternatives. The model will offer cranes managers a new tool to help them make decisions on how best to manage lands for the sandhill crane population.

The causes and possible effects of declining fat storage rates on midcontinent cranes are still under study. A decline in fat reserves would be expected to have greater adverse consequences for those cranes having the longest migrations and for those breeding in the most extreme environments, where food is least available early in the breeding season. However, based on body mass measurements from 1996, fat reserves of the smaller-bodied and northern-nesting lesser sandhill cranes may have been affected by changing habitat conditions in the Platte River Valley than have fat reserves of the larger subspecies, which breed farther south under less severe conditions.

Recent declines in the rate of fat storage in the Platte River Valley underscore the potential risk of species becoming inextricably linked to agriculture. Many species, like the sandhill crane, have become dependent on grain residues to meet their energy and fat storage needs during much of the year as native habitats that formerly supplied their needs have disappeared (Plate 7.5). For species deriving energy and fat storage needs from agricultural foods, the long-term outlook is uncertain because numerous factors can and are diminishing the availability of agricultural foods. Moreover, it is unlikely that agriculture will return to less efficient harvesting methods such as those which existed in the recent past. Thus, species such as the sandhill crane which have highly specialized roost-site requirements that limit their flexibility to adjust to changes in distribution and abundance of food are most likely to require human intervention to ensure needs are met.

Shrinking Channels, a Growing Problem

In the late 1970s, the midcontinent population of sandhill cranes roosted in about 51 miles (82 km) of the Platte River channel between Lexington and Chapman, a marked decline from the approximately 120 miles (193 km) of river available to cranes between the North Platte–South Platte confluence and Grand Island earlier in the century. By 1996, sandhill cranes were restricted to staging along a) approximately 40 miles (64 km) of the Platte River, b) the North Platte site near Sutherland, and c) a small area at the upper end of Lake McConaughy (see Map 7.1). Much of the Platte River to the west of Kearney has been abandoned or has limited use by cranes. Continued shrinkage of crane roosting habitat in the Central Platte Valley is causing crane density and competition to increase, contributing to foraging farther away from the river and higher energy costs. The disproportionate amount of time cranes spend searching for invertebrates in wet meadows and grasslands despite declining fat storage rates suggests that they are having difficulty meeting their protein and/or calcium needs under current conditions; indirectly, that difficulty may be contributing to the declining rate of fat storage.

Trends and scale of habitat change in the Platte River Valley in the past 60 years raise the concern that the midcontinent sandhill crane population and other waterbirds could lose this key staging area sometime in the 21st century (Plate 7.6). Such a loss would have eliminated the best site on the planet for viewing large numbers of cranes and would bring to an end the annual visits of thousands of birders who come from across the United States and from other countries to see the cranes and other wildlife. Loss of the Platte River staging area would have unknown but potentially severe adverse consequences for the midcontinent population of sandhill cranes, which depend on this ecosystem for a major part of their nutrient needs for spring migration and reproduction.

The Platte River has been a strategic staging area for migrant waterfowl in a part of the Central Flyway that has lost much of its wetland resources to agricultural development. In the Rainwater Basin Area (RBA), key Central Flyway waterfowl staging area adjacent to the Platte River Valley on the south, fewer than 400 of the original 4,000 wetland areas remain. As a result, crowding of waterfowl occurs in dry and unusually cold springs, and several hundred thousand ducks and geese have died due to epizootics of avian cholera during the past 25 years. In March 1998, an estimated 100,000 snow geese died from avian cholera in the RBA. To date, waterbird loss from disease in the RBA has been diminished because many Central Flyway waterfowl in dry and cold springs move to the Platte River Valley, where few disease problems have existed. Thus, changes that diminish wetland habitat on the Platte to waterfowl pose the specter of increased crowding and higher losses of waterfowl and other waterbirds from disease in the RBA.

Working toward a Solution

Efforts to protect remaining meadows and to maintain wide channels have become a high priority for conservation organizations and government agencies attempting to preserve the Platte as a key staging area for the sandhill crane and the endangered whooping crane. A section of the Platte River Valley 3 miles wide (about 5 km) and 56 miles long (90 km), from Lexington to Denman, was designated as critical habitat for the whooping crane under the Endangered Species Act of 1973. About 13,000 acres of the Platte River Valley have been acquired in fee title or perpetual easement for migratory waterbird use in tracts located from near Overton to south of Grand Island. Most of these lands have been purchased by the Platte River Trust, the Nature Conservancy, and the National Audubon Society. Conservation lands are being actively managed to meet the needs of sandhill and whooping cranes and other migratory waterbirds, including the least tern (classified as "endangered" in the Great Plains) and the piping plover (classified as "threatened"). Terns and plovers nest on sandbars in the Platte River.

To adequately protect the channels for roosting and the wet meadows for foraging, instream flows must be sufficient to maintain wide, shallow river channels and, in the surrounding meadows, high water tables. High-energy cereal grains are also crucial, but if roosting habitat and meadow habitat remain widely distributed, much more cropland is more accessible to cranes for obtaining corn to meet their energy and fat storage needs.

Securing adequate flows to meet needs of migratory waterbirds has been a difficult task because of intense competition for the Platte's waters. Historically, water in rivers of the Great Plains has been appropriated mostly for consumptive uses, particularly for irrigation and hydropower development. In line with the doctrine of appropriation, water rights have been allocated based on chronological order of application and, so long as water laws are followed, the water right is maintained indefinitely. At the Platte, about 70% of average annual flows were allocated before the need for adequate instream flows to maintain the riverine ecosystem was widely recognized.

The growing body of information about the habitat needs of sandhill cranes, whooping cranes, and other waterbirds using the Platte River Valley has helped ensure that allocation of flows for consumptive uses upstream have been appropriately mitigated in the past two decades. However, habitat losses from water projects completed before the maintenance of instream flows for wildlife became a consideration must also be addressed: a major part of habitat loss is from projects completed decades ago.

A big step toward ensuring that adequate habitat will be maintained for migratory waterbirds in the future was taken in July 1997 when a cooperative agreement was signed by the governors of Nebraska, Colorado, and Wyoming and by Interior Secretary Bruce Babbitt. The agreement lays out a comprehensive habitat restoration plan with a long-term goal of making available 130,000 to 150,000 additional acre-feet of water each year in the Platte during those periods when greater flows are most needed to meet migratory bird requirements. Also, the plan calls for restoration and protection of 29,000 acres of habitat between Lexington and Chapman, Nebraska, over the next 25 years. Habitat complexes are to contain those key elements that research has indicated are necessary to attract and meet the needs of whooping and sandhill cranes, including wide, unobstructed channels and large tracts of wet meadow and grassland. Once restored, each habitat complex would have the potential for attracting thousands of sandhill cranes, thereby reducing pressure on existing areas that are still inhabitable.

Although the July 1997 cooperative agreement is nonbinding, its failure would cause conflicts to resume that have led to decades of litigation and stalemate over how the waters of the Platte River Basin should be managed. Thus, while habitat conditions for numerous species of migratory waterbirds remain precarious in the Central Platte Valley at the approach of a new century, there is renewed. Progress, though slowed by a myriad of water-related issues, is continuing because of the strong commitment of many in the conservation community, government, and public to maintain this world-class gathering of birds. Driving this effort is a recognition that if the crane chorus falls silent along the Platte, one of nature's most enduring and spectacular rituals of spring will have been lost—and on our watch.

Beyond the immediate problems and potential solutions for sandhill cranes along the Platte, the long-term well-being of the midcontinent population of sandhill cranes will require a comprehensive understanding of the needs of cranes throughout the annual cycle. At the present time, we lack a thorough understanding of the habitat needs of specific subpopulations or even subspecies. We do not know the breeding grounds of subspecies staging along various sections of the Platte, their migration routes, where they spend the winter, or even how long they stay and draw resources from the Platte.

In March 1998, I had my technician, Dave Brandt, attach satellite-linked transmitters to 1 greater, 2 Canadian, and 2 lesser adult sandhill cranes in the Platte River Valley as a first step toward gaining a better understanding of where cranes of the three subspecies spend the rest of the year. Preliminary results from this study, which is scheduled to continue through the year 2001, are encouraging. Locations of the satellite-monitored cranes are relayed back to my laboratory every 4 to 7 days, providing types of information previously not available. As a result of this new technology, it will be possible to follow cranes throughout the annual cycle, pinpoint those areas which are most important to the birds, and put crane managers in a better position to focus attention where most needed.

Maintaining healthy populations of sandhill cranes and other migratory waterbirds in the face of continuing change in the American landscape is not easy: it will require the continued strong support of the public, particularly those who derive special enjoyment from watching birds. In the meantime, if you have not traveled to the Platte in late march and seen the cranes, I encourage you to do so. There is no similar gathering of cranes on this planet or beyond—except perchance in some far valley of the Milky Way.

References

1. Krapu, G.L., and J.L. Eldridge. 1984. Crane River. Nat. Hist. 93(1):68 75.
   A general overview of how upstream water development has affected the staging ecology of sandhill cranes in the Platte River Valley.
2. Krapu, G.L. 1996. The Platte River Ecology Study. Northern Prairie Science Center. http://www.npsc.gov/resource/othrdata/platteco/platteco.htm (19 Jan 1996).
   A semitechnical report summarizing the findings of a multidisciplinary research project undertaken to identify the habitat needs of sandhill cranes and several other waterbirds in the Platte River Valley.
3. Walkinshaw, L.H. 1973. Cranes of the World. New York: Winchester.
   A general summary of life history information on cranes of the world, this book was written for a general audience and contains numerous photos.
4. Williams, G.P. 1978. The Case of the Shrinking Channels—The North Platte and Platte Rivers. U.S. Geol. Surv. Circ. 781.
   A technical report describing the changing dimensions of the Platte River and including photos of the river before upstream water development reduced channel width.
5. Rollins, P.A., editor. 1935. The Discovery of the Oregon Trail—Robert Stuart's Narratives. New York: Scribner's.
   A historical narrative, part of which describes the explorer's visit to the Platte and North Platte rivers in the early 19th century.
6. Walkinshaw, L.H. 1949. The Sandhill Cranes. Cranbrook Inst. Sci. Bull. 29.
   A general summary of life history information on sandhill cranes.
7. Krapu, G.L., D.E. Facey, E.K. Fritzell, and D.H. Johnson. 1984. Habitat use by migrant sandhill cranes in Nebraska. J. Wildlife Mgmt. 48:407–417.
   A technical article describing habitat preferences of sandhill cranes during spring in the Platte River Valley; based on studies employing radiotelemetry.

8

BRIAN A. HARRINGTON

The Hemispheric Globetrotting of the White-rumped Sandpiper

As a group the shore-birds cover greater distances in the aggregate in their spring and autumn flights than any other birds. . . . It is strange to find the white-rumped sandpiper, a species known well to few ornithologists in the United States, the most abundant of the wintering shore-birds on the pampas. —Alexander Wetmore, *The Migrations of Birds,* 1926

The white-rumped sandpiper, as becomes clear from Brian Harrington's chapter, is one of our more obscure shorebird species. Imagine that in 1998 the details of where it goes during a major portion of its migration are still unknown! In many ways, however, it is a showcase migrant, for it exhibits in extreme form some of the hallmark characteristics of long-distance migration. Timing is everything. In early June I have seen flocks of white-rumps feeding in wet fields in southwestern Kentucky. Presumably most were females, that sex arriving later on the breeding grounds. Nonetheless, these same birds would be incubating eggs on the highest Arctic tundra within two weeks of when I watched them, having in the meantime migrated some 2,000 miles, been courted by males, built a nest (such as it is), and completed a clutch of four eggs at a rate of one egg per day. White-rumped sandpipers spend far more of their lives in transit than they do at any one place. Any individual white-rump is probably migrating between breeding and wintering areas for at least six months of every year. The white-rump illustrates, perhaps better than any other bird described in this book, the complex challenges faced by long-distance migrants: flying different routes northward and southward, utilizing very different habitats during the course of a year, and feeding on different prey species. That such a quintessential migrant has only now come under close study provides ample testimony to just how much remains to be learned about the migration of even relatively common species. — *K.P.A.*

How is it possible that the white-rumped sandpiper, barely larger than a sparrow, would come to have a migration that spans the extremes of North and South America? And why does its migration include flying—without stops—thousands of miles over the ocean? The preparation and endurance needed for the flight are awesome. They include laying on masses of fat, enough to almost double their weight, in just two weeks before the flights. They also include flying at altitudes where mountaineers have difficulty getting sufficient oxygen. These and innumerable other questions provoke my amazement and fascination for a relatively little known shorebird—the white-rumped sandpiper—one of the small "peep" sandpipers that is unique to the Western Hemisphere.

White-rumped sandpipers undertake migrations that are not only amazing but that also challenge the assumptions we easily make about the inherent risks of migration. As a scientist, I wonder what advantage is gained by traveling to Argentina or Chile instead of Texas, Florida, or Georgia? Or why don't shorebirds just stay in southern South America the year-round? (Some "austral" species of shorebirds do just that.) Why do migrations that span tens of thousands of miles improve survival? The list of questions goes on. This account treats migration itself, using information about the white-rumped sandpiper and contrasting it to a close relative, the Baird's sandpiper.

White-rumped sandpipers are one of the so-called *peeps*, a group of small, similar-appearing sandpiper species that in North America include sanderling, dunlin, least, semipalmated, western, and Baird's sandpipers. Although these species may appear similar in casual view, each has reasonably distinct characteristics that are well described in good field guides. White-rumped sandpipers are perhaps closer in size and appearance to the Baird's sandpiper than to any of the others. Both breed and spend their winters at similar northern and southern latitudes, yet they have surprisingly different migration patterns. We will return to this contrast later.

Breeding Activities

The summer arrival of shorebirds in the Arctic is well described by Roland Clement, who said, "Yesterday there were no birds; today they are everywhere." White-rumped sandpipers evidently return directly to breeding areas from staging sites in the United States. Arrival is typically in early June when snow may still cover swales and the north sides of tundra hillocks and snowfall is still likely. Nevertheless, breeding activities begin almost immediately. William Drury saw flight and song displays of white-rumped sandpipers on the first day (19 June) that the sandpipers returned to Bylot Island, where he studied their nesting behavior through the summer [1]. (Bylot Island is the easternmost place this species is known to nest, and the arrival date there is later than in the main part of the range.) By back-dating a clutch of eggs he found later, Drury estimated that the last egg had been laid on the 25th or 26th of June; white-rumped sandpipers lay 1 egg about every 30 hours [2], so the first had apparently been laid just a few days after the sandpipers had returned (Plate 8.1). This timing is remarkable when we consider that most of the white-rumped sandpipers pass north through the central United States during the end of May or early June [3] and that their arrival dates in the main parts of the breeding range are only a few days later [4]. Thus, just before beginning their frenetic breeding and territorial displays, there is virtually no time to replenish the reserves used up during the flight from the central United States to areas north of the Arctic Circle. Then, almost immediately, the female begins to produce a clutch of 4 eggs that will equal about 90% of her body weight [5]. The energetic demands met by these small feathered bodies, as they pitch into Arctic ecosystems at the end of a flight exceeding 10,000 miles, are enormous.

There is a growing consensus among biologists that for Arctic sandpipers there is a premium placed on reaching breeding grounds at the earliest possible stage (but not so early as to freeze or to starve to death) in order to secure the best possible territories [6].

White-rumped sandpipers have a *polygynous mating system* (males attempt to attract and mate with more than one female) and the males take no part in the care of nests or young [2]. It is likely that a female uses the quality of the territory a male has secured (successfully defending it against other males) as an important factor in her choice of a mate. This selection process, in turn, suggests that there is competition among females for the males, thus putting a premium on their early arrival on the breeding grounds as well; in fact, female white-rumped sandpipers arrive at about the same time as males [4].

Like most kinds of shorebirds, white-rumped sandpipers breed in open tundra habitats where vision as well as sound play important communication roles. Elaborate aerobatic displays and vocalizations are key to male sandpipers' mating success. Certain displays advertise to prospective mates, while others proclaim territorial ownership. Performances may be at any time of the day (above the Arctic Circle, daylights exists almost 24 hours a day during June). Increased altitude will enhance detection of displays across the expansive tundra habitats, so it is not surprising that aerial performances (well described in Drury's study) are integral to pairing and territorial activities. Edward Miller points out that the display height of shorebirds tends to vary inversely with the sizes of territories and the distances between pairs [7]. However, at the same time, long broadcast distances set limits to the structure of displays, which need to be obvious, stereotyped, and redundant for most effective reception by females. Presumably the most effective male suitors strike the balance most attractive to potential mates, but the equations for success will vary with terrain, weather, and other factors. Whatever the case, it is clear from Drury's account that male performances are intense and require a huge investment of time and energy. The principal sound of white-rumped sandpipers in the aerial flights is a prolonged, rattling buzz, sounding like a running fishing reel (or the carriage return of a typewriter) interspersed with a disyllabic "ng-ock" note, likened to the sound of a small pig [1].

A male defends his territory vigorously until clutches are complete, or in other words for perhaps one week; during this time, his tenancy will likely be challenged repeatedly, sometimes involving vigorous fights with other males [1]. Males keep a steady guard on their territory; at times there may be few chances to hunt for food, which that early in the season is probably scarce anyway. It is unclear whether successive females that mate with one male use nests in his territory or elsewhere, but sometimes nests can be surprisingly close together, just 12 m apart in one instance [4]. Males leave the care of eggs and young entirely to the females.

Southward Migration

We have little information on the activities of the male white-rumped sandpiper after mating, but the males evidently leave the breeding grounds well before the females. W.E.C. Todd notes that the white-rumped sandpiper is a very early migrant on the coast of Labrador [8]. At the Magdalen Islands, near the mouth of the Gulf of Saint Lawrence, Raymond McNeil and Françoise Cadieux found that the sex ratio was about equal among the earliest-arriving contingents of white-rumped sandpipers during late July [9], perhaps because the earliest migrants were failed breeders and/or nonbreeders. By August the males noticeably outnumbered females, and this disparity remained until the researchers ended their studies early in September.

Juvenile white-rumped sandpipers are among the last northern sandpipers to begin the southbound migration. George Miksch Sutton found flocks of juveniles (and one confirmed adult) remaining well after freeze-up had begun at Southampton Island, in northwestern Hudson Bay, during early October [10].

The evidence that white-rumped and other small sandpipers are flying directly over the ocean between Canada and South America is clear. In the Magdalen Islands, in the Gulf of Saint Lawrence, numbers of adult white-rumped sandpipers decline during late August and early September, which is when flocks of adults begin to appear in Surinam on the northeast coast of South America [3]. There is no passage of appreciable numbers of white-rumped sandpipers on the U.S. Atlantic coast at this time; and virtually none of the many hundreds that have been marked with dye during July and August in the Magdalen Islands and on the southwest James Bay coast have been found in the United States.

The distance from the Magdalen Islands to the Surinam coast is about 2,900 miles. If the white-rumped sandpipers were flying at ground speeds of 40 to 45 mph, a reasonable estimate, they would require 65 to 75 hours to complete the flight, assuming no assistance or hindrance from winds [3]. But if they are like most other shorebirds, they launch their flights when assistance can be gained from weather systems, especially tail winds that come with cold fronts arriving from the northwest [11].

Biologists also think we can roughly estimate how far a sandpiper can fly if we know its size (wing length), its weight, its flight speed, and how much fat it has accumulated before departure [12]. A white-rumped sandpiper with little fat might weigh 30 to 40 g (less than 1.5 oz), but a very fat one might weigh 60 to 65 g. According to the energetics equations used for estimating shorebird flight capacities, 20 g of fat would be sufficient to sustain flight of a white-rumped sandpiper for roughly 40 to 45 hours [3]. Using simple arithmetic, you will probably conclude that the sandpipers fall into the ocean between 500 and 1,000 miles from the South American coast. Yet, tens of thousands do arrive on the coast of Surinam each autumn. Sometimes we scientists are humbled by the critters we study. A key factor not included in the above estimates is assistance from favorable winds. It also is possible that the formulas for estimating flight energetics are too conservative. Research in which flight energetics has been more directly measured on some larger oceanic birds has shown that the energetic cost of flight was much lower than our commonly used formulas would project [13].

I expect readers are thinking about some of the complexities that might be involved in what I have just described. How can white-rumped sandpipers almost double their weight in a week or 10 days? If *I* were to accomplish that, how could I then undertake a grueling marathon that would cover 2,500 miles without stops, food, or water and that would last at least 3 or 4 days? And there are questions about flight itself. To become airborne a bird needs to achieve lift from the wing. There is a complex relationship between wing surface area, weight, and speed—all feeding into the bottom line of successful flight. But this balance shifts to an extraordinary degree when the mass or weight of the bird changes. Added to this is the fact that the sandpipers will travel at high altitudes, sometimes exceeding 15,000 feet. As with aircraft, high altitude achieves some efficiencies through reduced drag, but unlike aircraft the birds must operate within the constraints of a metabolic system that requires large amounts of oxygen, which is less available at higher altitudes. At these altitudes, moreover, the air is "thinner," meaning that the forces of lift acting on the wing are once again reduced. Aircraft can alter that equation by increasing speed, sometimes beyond the speed of sound itself. This change of speed forces more air to pass over the wing, increasing lift. Birds can increase their speed only so much.

On the Surinam coast the white-rumped sandpiper's migration route merges with the routes of a variety of other shorebirds, such as yellowlegs, dowitchers, willets, and semipalmated sandpipers. For these species, in contrast to the white-rumped sandpiper, the major stopover sites in North America are south of Labrador and the Gulf of Saint Lawrence, meaning that their final flight to South America is shorter (Map 8.1).

The broad tidal flats characteristic of the Surinam coast consist of a soupy mud known as sling mud. When I visited in 1982 I saw a small boat actually sail (and make headway!) on the mud far back from the low-tide line. My first attempt to walk on the mud (to get a better look at a flock of sandpipers) quickly showed me what every local resident knows—you can't walk on the mud, or make any headway short of somehow swimming. Inland of the shoreline is mostly mangrove swamp, a band several miles wide along much of the coast, making it very difficult to access the coast. To get a better perspective, my co-worker, Linda Leddy, and I hired a small aircraft on 23 August to fly low along the coast in an effort to find where the shorebirds were located. During the first 15 minutes of our survey we estimated we saw more than 42,000 shorebirds. More than 35,000 were peep sandpipers, including small numbers of white-rumps which had probably arrived in Surinam during the preceding three or four days.

A little more than a week later we had made our way to the coast by boat near the mouth of the Surinam River. We arrived just at dusk at a very short section of coast where there was enough high ground for the Surinam Forest Service to build a small visitor camp. Soon after dawn the next morning I watched flock after flock of sandpipers, many of which were white-rumps, passing east along the coast toward Brazil. Counts I kept showed that birds were passing my position at a rate of about 4,500 per hour, but by 10:00 a.m., when a fairly stiff head wind had developed, the passage had ended. A similar flight happened during the next two mornings that we were at

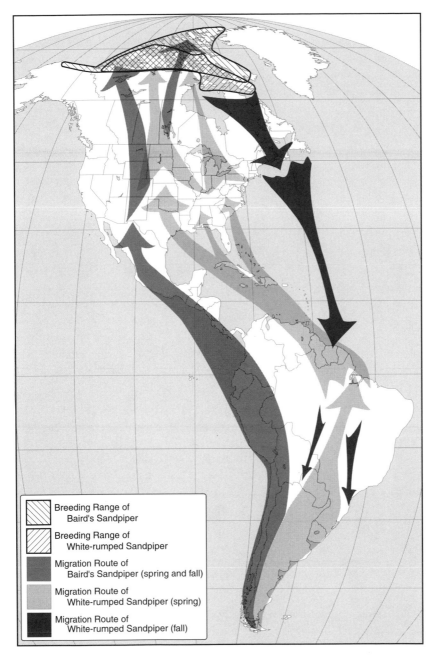

Map 8.1 The breeding ranges and migration routes of the white-rumped sandpiper and Baird's sandpiper.

Matapica, and roughly 10% of the peeps were white-rumped sandpipers. Later in the day I was able to get close enough to two feeding flocks to see that all the white-rumped sandpipers were adult birds. None of them appeared to be very fat. In fact, none of the birds caught by Arie Spaans, who spent three years studying the shorebird migration in Surinam during the 1970s, had enough fat to fly any great distance [14]. Spaans also found that the white-rumped sandpiper passage is mostly completed in Surinam by the end of September.

At this juncture we lose track of the southward white-rumped sandpiper migration, but fragments of information suggest they continue east along the coast to the northern Brazil coast. About a month elapses before white-rumped sandpiper numbers start to increase in the state of Rio Grande do Sul, on the southernmost coast of Brazil near the border with Uruguay [3]. Meanwhile, there is no passage of white-rumped sandpipers in the state of Pernambuco in easternmost Brazil, where routine shorebird census and banding research were conducted year-round by Severino Mendez and co-workers over a 1 ¼-year period [15]. They never found white-rumped sandpipers at any season.

Taken together, the available circumstantial information argues strongly that the white-rumped sandpipers fly overland from the north coast to the south coast of this huge nation, a distance of roughly 2,200 miles. In his 1983 consideration of shorebird migration in Brazil, Paulo Antas suggested that white-rumped sandpipers follow a Central Brazil flyway which begins near the mouth of the Amazon River and follows some of the major river valleys that run north–south into Uruguay and Argentina [16]. It is uncertain whether the white-rumped sandpipers typically traverse this distance in nonstop flights, or whether they stop to use the sandbars and mudflats that characterize these rivers during October and November, when the rivers still are low. In fact, there is little question that many birds do stop along the way, but it is unclear whether these represent fallout from an overhead passage or whether it is their normal strategy. In any case, the passage of the migration swells in southern Brazil and more so in northern Argentina during September and October and through November.

Boreal Winter Period

Although some individuals favor nonmarine habitats [17], most white-rumped sandpipers are largely marine shorebirds during their winter period [18] (Plate 8.2). (In deference to my Chilean and Argentine friends, and to the reality of the situation, I should explain that my use of the term "winter" here refers to the birds' winter period, even though it is summer where they winter.) During the austral spring and early summer, marine invertebrate populations have doubtless grown into a seasonally predictable largess for white-rumped sandpipers to feast upon. In addition, day lengths are at their longest of the year during December, giving more daylight time for foraging. Finally, the tidal amplitudes exceed 20 feet along much of the Argentine and Chilean coast, which of course provides excellent conditions for birds of the intertidal zone. On the other hand, at these latitudes during June (austral winter) there are only eight hours between sunrise and sunset. Therefore, there would not be good daylight foraging conditions during low tides on about 25% of the days (nocturnal foraging has not been studied in white-rumped sandpipers, but in other shorebirds where it has been studied it apparently is substantially less productive [19] than daytime foraging). So, it is reasonable to think that this would not be a good place for intertidal foragers during the austral winter.

The daily routines of white-rumped sandpipers in Argentina typically include a) foraging whenever tidal conditions permit and b) resting in large flocks, often comprising thousands of birds of several species, at other times. Some individual white-rumps return to the same wintering locations year after year, where they may defend territories; those which forage on the tidal flats tend to be nonterritorial, whereas those in the nonmarine wetlands or along tidal sloughs may defend territories [17].

Bird feathers can be divided into two major categories, those specialized for assisting flight (*flight feathers*) and all others (*body feathers*). Even with care, feathers wear out; birds periodically need to molt and grow new feathers. Although the keratin which forms feathers is a resilient material, it is constantly subjected to abrasion and wear; unlike human fingernails, bird feathers do not grow continuously but instead become inanimate once they reach full size. Thereafter, feathers gradually wear and become frayed, sometimes are chewed by feather lice, and/or become brittle and faded from constant harsh exposure to weather and sunlight. Shedding old feathers and growing new feathers is essential to birds for maintaining good condition, but it also requires a substantial energetic investment [20]. Consequently, birds generally do not molt while breeding or migrating because the high energy requirements are mutually exclusive.

Most kinds of birds molt flight feathers soon after the breeding season. In general, there is a complete molt of the flight and body feathers before the southward migration begins. White-rumped sandpipers and many other Arctic shorebirds are exceptions to this pattern; although body feather replacement may begin soon after breeding, the flight feather molt does not start until the southward migration is completed.

Arctic shorebird molt is evidently deferred because the birds need to migrate south as soon as possible after breeding. There is ongoing debate among ornithologists as to why this is so. One idea is that the Arctic summer is simply too short to leave time for molting on the nesting grounds. David Schneider and I looked at another idea. Studying the invertebrate prey consumed by shorebirds, we showed that there was a substantial depletion of that prey at the migration staging site [21]. This depletion was caused by the shorebirds themselves, which suggests that the earliest-arriving birds may fare better in fattening up for their continued migration than later-arriving ones. Maybe there is early migration in shorebirds because natural selection has favored the early birds that catch the worms!

The flight feather molt of white-rumped sandpipers has never been studied, but it is likely to be similar to that found in other shorebirds that migrate to similar latitudes [22], probably requiring somewhere between 60 and 90 days to complete. So individuals starting the molt in mid-November might not finish until mid-January or February. During this time the weight of the sandpipers remains low [3], and that is probably is no accident: attempting to fly with missing wing feathers and with extra weight might not work well. Shorebirds that are not nimble in flight are also more vulnerable to raptor predation [23], a very important factor in shorebird survival.

Northward Migration

Some white-rumped sandpipers begin to leave their winter quarters as early as the middle of February, but it is unclear whether this early departure is related to changing habitat conditions locally or to a gradual start of the northward migration. When working at Peninsula Valdez, on the central Atlantic coast of Argentina, my co-workers and I found a rapid decline in numbers of white-rumped sandpipers during late March. We had not caught any, so we did not know whether they had accumulated fat prior to leaving this part of the argentine coast. But Frans Leeuwenberg and his wife, Suzana Lara Resende, were catching and weighing white-rumped sandpipers at this same time of year in Rio Grande do Sul, in southern Brazil (Plate 8.3). It is unclear where these earliest-departing birds were going, but evidently they were not about to travel a long distance as they had not gained much fat [3]. By late April, however, the white-rumped sandpipers at Lagoa do Peixe were gaining fat, and during the first week of May some of them were exceeding 60 g—some of the heaviest weights we have found for them, and almost double the weight of the lightest individuals during early April. Many of the birds had sufficient fat to be able to fly 2,500 miles according to current formulas.

Evidently, white-rumped sandpipers travel in a rapid, nonstop flight between southern and northern South America. During the northward migration the water levels are very high on the rivers that pass north–south through central Brazil [16], exposing very few riverbanks and sandbars, unlike the situation during the southbound flight.

Where white-rumped sandpipers go in northern South America remains something of a mystery. Groups of hundreds often are found on the coasts of Surinam and Venezuela, but these are much smaller numbers than occur in northeastern South America during the southward migration. Raymond McNeil and co-workers captured a small number near the Sucre Peninsula in eastern Venezuela, but the weights were relatively low which suggests that the area was not being used as a migration staging area [3]. Similarly, the weights were low in samples caught by Betsy Trent Thomas in the inland llanos of Venezuela [3]. Wherever the white-rumped sandpipers go, they evidently do not linger for very long, because by the end of April they are beginning to appear in the midwestern grasslands of the United States; remarkably, this is roughly the same time that we had recorded the major departures from southernmost Brazil in 1985 [3].

By making routine counts at Cheyenne Bottoms through the spring, Ed Martinez found that numbers of white-rumped sandpipers began to build up in Kansas during late April, and continued to increase steadily until the last week of May or the first week of June, at which time the numbers drop suddenly [3] (Plate 8.4). It is clear that the white-rumped sandpipers visiting Cheyenne Bottoms were staging there or, in other words, that they remained there and steadily gained fat reserves; by the time of the departure many of them weighed more than 50 g, having gained enough fat to easily fuel nonstop flights of 1,800 to 2,000 miles into the Arctic breeding range, where they typically arrive during the last week of May or the first week of June.

We have now come full circle, so to speak, in our tracking of an ounce-and-a-half package of bone, feather, and protein between the continental extremes of the Western Hemisphere. From the perspective of our human domain, it seems absolutely incredible that such small birds would undertake these amazing migrations, that their life-style must be fraught with high risk and frequent deaths. In our world it is unthinkable to set off on a 2,000-mile journey across ocean waters without special vehicles, or without extra larder in our baggage. But if we stop to consider evolutionary theory, we need to rethink this perception: there must be some advantage gained by the birds whose seasonal homes shift so dramatically. Consider, then, what is happening in the course of the white-rumped sandpiper's year. Each spring they arrive in the Arctic just as the spring/summer burst of life is coming into full seasonal blossom. The resources that quickly become available include a profusion of insects, spiders, and other invertebrates. Because the winter season is so harsh that only very small populations of other animals (and very few different species) survive in the Arctic, there are few "resident" competitors for the summer resources. In short, there is a largess available to all comers who can get there to take advantage of the seasonal abundance.

In the next step of their year the sandpipers travel south and east to the marine coasts of the United States and Canada. Here they fatten primarily on intertidal invertebrate animals. Studies of these communities in the northeastern United States and Canada [24] show a seasonal pulse of the marine invertebrate animal populations occurring during July and August, just when the sandpipers arrive to begin reaping that predictable larder.

The next migration stepping-stone is in northeastern South America, especially the coast north and west of the mouth of the Amazon River. Given that temperatures are fairly steady throughout the year in this region, we do not think of the climate there as a highly seasonal. With respect to the intertidal animal community, though, it probably is highly seasonal. The sling mud along the Surinam coast, mentioned earlier, has its source in the Amazon River. Throughout the immense Amazon watershed, there are dry and rainy seasons with especially large rainfalls occurring in the upper reaches. The volume of water that comes down the Amazon is so huge that downriver water levels commonly fluctuate by 30 to 40 feet between seasons. Huge seasonal pulses of fresh water flow from the Amazon into the Atlantic and get carried north and west by the Guiana Current toward the coast of Surinam [25]. There have not been comprehensive studies of the effect of these seasonal salinity changes to the invertebrate animal populations, but preliminary information suggests that seasonal blooms of the invertebrates occur during September and October, when the shorebirds are visiting.

The next major zone for the white-rumped sandpipers is in austral South America. Once again, there is a very pronounced seasonal pulse: the highest wetland and coastal invertebrate population growth occurs during the austral summer, when the boreal shorebirds are visiting.

On the flight north we can point to another predictable larder for the white-rumped sandpipers when they visit wetlands in the midwestern United States and southern Canada during April and May. This is spring, the time when there is a huge production of invertebrate animals in the wetlands, ranging from great numbers of midges and other insect larvae to copepods and other crustaceans in the water.

We often think of shorebird migrations as having developed as an escape from harsh, northern-winter climates; but they may have had very different evolutionary roots. Rather than simple escape from a hostile environment, the strategy may be to move from one seasonally abundant resource to the next throughout the entire year. The *stepping-stone pattern* we see in the white-rumped sandpiper's migration cycle appears to be a highly refined itinerary which takes them—by a series of long-distance, nonstop flights—between predictable, but seasonally ephemeral, food resources scattered at distant points of the hemisphere. Thus, the migration pattern we see may have evolved as a way of linking together seasonally rich food resources at widely separated points of the globe.

To consider this last idea briefly, let's compare the migration of our white-

Plate 8.1 A white-rumped sandpiper on its nest in the High Arctic. (Photo by J.P. Myers/Vireo)

Plate 8.3 In the open environments where shorebirds live, capturing the migrants in mist nets can be a difficult task. (Photo by D.C. Twichell)

Plate 8.2 Tidal flats on the Argentine coast used by white-rumped sandpipers during "winter." (Photo by D.C. Twichell)

Plate 8.5 Baird's sandpiper in breeding plumage. Superficially similar to the white-rumped sandpiper, and sharing its breeding and winter range, the Baird's migration route is very different. (Photo by J.P. Myers/Vireo)

Plate 8.4 White-rumped sandpipers in spring are rather more colorful than they are in autumn. (Photo by A. Morris/Vireo)

Plate 9.1 The swarms of dunlin at the Copper River Delta, with the Chugach Mountains as a backdrop, are a magnificent sight. (Photo by J.P. Myers/Vireo)

Plate 9.2 Western sandpipers and dunlins form huge mixed flocks, as seen here, in both spring and fall. (Photo by J.P. Myers/Vireo)

Plate 9.3 A spring migrant western sandpiper caught in a mist net. (Photo by M.A. Bishop)

Plate 9.5 Black turnstones (shown here) and other shorebird species stage for migration at sites near the Copper River Delta. (Photo by K.P. Able)

Plate 9.4 This spring migrant western sandpiper has been outfitted with a small radio transmitter, glued to its back. Note the thin wire antenna extending beyond the bird's tail. (Photo by David Weintraub)

Plate 10.1 A female rufous hummingbird feeding from and pollinating red columbine at Elfin Cove, Alaska. (Photo by W.A. Calder)

Plate 10.2 Breeding broad-tailed hummingbirds (like the one shown here) compete with migrant rufous and calliope hummingbirds for access to flowers. (Photo by D. True/Vireo)

Plate 10.3 A silver mine, abandoned without reclamation of tailings and slag heaps, is finally being revegetated by *Corydalis,* which depends upon transient rufous and calliope hummingbirds for pollination. (Photo by W.A. Calder)

Plate 10.4 Hummingbird banding at Rocky Mountain Biological Laboratory, Gothic, Colorado. (*Top*) Barry Rowe removes a rufous hummingbird from a mist net. (*Bottom*) A male rufous hummingbird awaits weighing and banding. (Photos by W.A. Calder)

Plate 10.5 Rufous hummingbirds are stoically tranquil when handled for banding and weighing, often continuing to lie calmly when free to go. Birds are weighed with the spring balance shown here. (Photo by W.A. Calder)

Plate 10.6 This young male rufous hummingbird did not find nectar in time. He perched under a rocky overhang where he slipped into a final bout of torpor to conserve his last blood sugar. After death, he became mummified in the dry air of the Santa Catalina Mountains near Tucson, Arizona. (Photo by W.A. Calder)

Plate E.1 Many migrating cattle egrets that make the mistake of landing on the Dry Tortugas, Florida, die of starvation and dehydration. (Photo by K.P. Able)

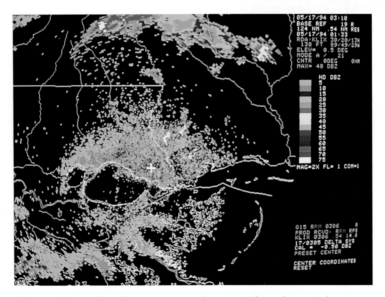

Plate E.2 The exodus of spring nocturnal migrants from the coastal areas of Louisiana and Mississippi on the evening of 16 May 1994, as seen by the WSR-88D weather radar at Slidell, Louisiana (near New Orleans). The green and yellow areas northeast of the radar station (marked with the +) show greater densities of birds emerging from mature bottomland forests. Many birds are also departing from cheniers and small forest patches along the coast and in the delta of the Mississippi River. Radar data of this sort can be used to identify hot spots where migrants consistently stop over in large numbers. (Photo by S.A. Gauthreaux Jr.)

rumped sandpiper to the very similar-appearing Baird's sandpipers [26] (see Map 8.1). These two species share common breeding and wintering ranges, but their migration "systems" are remarkably different. Although they often use wetlands, Baird's sandpiper will typically be found on the upper edges of wetlands or sometimes in grassy habitats. They find most of their food by spying it — in contrast to the white-rumped sandpipers, which typically probe for their food. Baird's sandpiper (Plate 8.5) is not a marine shorebird, as the white-rumped sandpiper is during much of its year. The route of Baird's sandpiper between its boreal breeding and austral nonbreeding range roughly follows the mountain and foothill cordillera of the Rocky Mountains in North America, and the Andes in South America, during both the northward and southward migration. Stopover areas include such habitats as the wetlands of the High Andes, or the playa lakes and other wetlands of the Rockies. Their migrations include some long, nonstop flights of distances comparable to those flown by the white-rumped sandpiper, but the types and locations of the stopovers they use are markedly different, evidently reflecting the different types of food they consume.

In contrast to the Baird's sandpiper, white-rumps use marine habitats during the southward migration. Thus their route deviates eastward of the route used for northward migration, passing through marine habitats of eastern North America where Baird's sandpipers are rarely seen. During northward migration, when white-rumps use mainly nonmarine habitats while in North America [3], they sometimes use the same stopover areas as Baird's sandpipers, but typically about a month later.

Yet after all these differences, the migrations of the Baird's sandpipers and the white-rumped sandpipers start and end in similar places, reflecting their globetrotting between the seasonally predictable but ephemeral resources of the Western Hemisphere which is their home.

References

1. Drury, W.H. Jr. 1961. The breeding biology of shorebirds on Bylot Island, Northwest Territories, Canada. Auk 78:176–219.

   Detailed description of observations made on nesting Arctic shorebirds.
2. Parmelee, D.F., and S.D. Macdonald. 1960. The Birds of West-Central Ellesmere Island and Adjacent Areas. Nat. Mus. Can. Bull. 169. Series 63.

   An annotated list with observations on the birds of Ellesmere Island.
3. Harrington, B.A., F.J. Leeuwenberg, S. Lara Resende, and R. McNeil. 1991. Migration and mass change of white-rumped sandpipers in North and South America. Wilson Bull. 103:621–636.

   A technical paper examining the energetics of the white-rumped sandpiper's autumn migration.

4. Parmelee, D.F., D.W. Greiner, and W.D. Graul. 1968. Summer schedule and breeding biology of the white-rumped sandpiper in the central Canadian Arctic. Wilson Bull. 80:5–29.
   Excellent description of the breeding-season life history of the white-rumped sandpiper.
5. Manning, T.H., and A.H. Macpherson. 1961. A biological investigation of Prince of Wales Island, N.W.T. Trans. Royal Can. Inst. 33(part 2):116–239.
   An account of biological exploration in the far north.
6. Myers, J.P. 1981. Cross-seasonal interactions in the evolution of sandpiper social systems. Behav. Ecol. Sociobiol. 8:195–202.
   A technical paper discussing the evolution of sandpiper social behavior.
7. Miller, E.H. 1983. The structure of aerial displays in three species of calidridinae (Scolopacidae). Auk 100:440–449.
   An analysis of the breeding-season flight displays of shorebirds.
8. Todd, W.E.C. 1963. Birds of the Labrador Peninsula. Toronto: Univ. Toronto Pr.
   Distribution and abundance of Labrador birds.
9. McNeil, R. and F. Cadieux. 1972. Fat content and flight range of some adult spring and fall migrant North American shorebirds in relation to migration routes on the Atlantic. Naturaliste Canadien 99:589–606.
   A technical paper discussing fat loads and flight distances as they relate to long overwater flights.
10. Sutton, G.M. 1932. The birds of Southampton Island. Mem. Carnegie Mus. 12:1–275.
    The results of ornithological explorations by a pioneer naturalist.
11. Richardson, W.J. 1979. Southeastward shorebird migration over Nova Scotia and New Brunswick in autumn: a radar study. Can. J. Zool. 57:107–124.
    A radar study of transatlantic shorebird migration and the conditions under which it occurs.
12. Pennycuick, C.J. 1989. Bird flight performance: a practical calculation manual. Oxford: Oxford Univ. Pr.
    An excellent introduction to the application of aerodynamic theory to bird flight.
13. Flint, E.N., and G.A. Nagy. 1984. Flight energetics of free-living sooty terns. Auk 101:288–294.
    A technical paper containing data on flight energetics in wild birds; uses the doubly labeled water technique.
14. Spaans, A.L. 1978. Status and numerical fluctuations of some North American waders along the Surinam coast. Wilson Bull. 90:60–83.
    A study of shorebirds at stopover and wintering areas in Surinam.
15. Junior, S.M. de Azevedo. 1992. Anilhamento de aves migratórias na Caroa do Avião, Igarassu, Pernambuco, Brasil. Cad. Omega Univ. Fed. Rural PE., Ser. Ci. Aquat., Recife. No. 3, p. 31–47.
16. Antas, P. de T.Z. 1983. Migration of Nearctic shorebirds (Charadriidae and Scolopacidae) in Brasil—flyways and their different seasonal use. Wader Study Bull. 39:52–56.
    One of a small number of studies concerning shorebird migration on the South American continent.
17. Morrison, R.I.G., and R.K. Ross. 1989. Atlas of Nearctic Shorebirds on the Coast of South America. Can. Wildlife Serv. Spec. Publ. Volume 1.

Provides much information on the distribution and abundance of shorebirds in South America.

18. Myers, J.P., and L.P. Myers. 1979. Shorebirds of coastal Buenos Aires Province, Argentina. Ibis 121:186–200.

    Studies of the wintering ground.

19. Turpie, J.K., and P.A. Hockey. 1993. Comparative diurnal and nocturnal foraging behaviour and energy intake of premigratory grey plovers *Pluvialis squatarola* and whimbrels *Numenius phaeopus* in South Africa. Ibis 135:156–165.

    A technical paper describing the foraging behavior of black-bellied plovers and whimbrels at their wintering areas in Africa.

20. Dolnik, V.R., and V.M. Gavrilov. 1979. Bioenergetics of molt in the chaffinch (*Fringilla coelebs*). Auk 96:253–264.

    A technical paper that analyzes the energetic costs of molt in a European songbird.

21. Schneider, D.C., and B.A. Harrington. 1981. Timing of shorebird migration in relation to prey depletion. Auk 98:197–220.

    A technical paper that examines the extent to which foraging migrant shorebirds deplete available prey for subsequent arrivals at migration stopover sites.

22. Thomas, D.G. 1971. Moult of the curlew sandpiper in relation to its annual cycles. Emu 71:153–158.

    A technical paper based on studies of the wintering grounds in Australia.

23. Marks, J.S. 1993. Molt of bristle-thighed curlews in the northwest Hawaiian Islands. Auk 110:573–587.

    A technical paper on one of the longest-distance shorebird migrants.

24. Whitlatch, R.B. 1977. Seasonal changes in the community structure of the macrobenthos inhabiting the intertidal sand and mud flats of Barnstable Harbor, Massachusetts. Biol. Bull. 152:275–294.

    A technical paper describing studies of the prey organisms fed upon by migrating shorebirds.

25. Duiven, C.S., and A.L. Spaans. 1982. Numerical density and biomass of macrobenthic animals living in the intertidal zone of Surinam, South America. Netherlands J. Sea Res. 15:406–418.

    Studies of prey organisms in the stopover areas of white-rumped sandpipers.

26. Jehl, J.R. 1979. The autumnal migration of Baird's sandpiper. Stud. Avian Biol. 2:55–68.

    A good summary of what is known about the migration of this relatively little studied species.

9

STANLEY E. SENNER

*Converging
North: Dunlins
and Western
Sandpipers
on the Copper
River Delta*

The restlessness of shorebirds, their kinship with the distance and swift seasons, the wistful signal of their voices down the long coastlines of the world make them, for me, the most affecting of wild creatures. I think of them as birds of wind, as "wind birds." —Peter Matthiessen, *The Shore-birds of North America,* 1967

My formative years, during which I became interested in birds, were spent in the landlocked Midwest. Shorebirds were things that I read about, not things that I saw very often. When later my family moved to western New York, I had my first real taste of shorebird migration, albeit still inland. Nonetheless, the spring flights of whimbrels along the north shore of Lake Erie can be very impressive. During my first spring there I saw flocks total-ing hundreds of individuals of a species that until that time had been only a fantastic resident of field guide plates. I thought this must be one of the great shorebird migrations of North America, but of course I was naive. Take a look at Plate 9.1 and see what a real shorebird staging area looks like. Places like this and Grays Harbor in Washington, the Bay of Fundy, Delaware Bay, and San Francisco Bay are now well known. As Stan Senner describes, even the remote Copper River Delta has become a mecca for en-thusiasts who want to see these remarkable spectacles — not for the purpose of seeing how many species they can tally, nor to search for some single rare stint in the swarm, but simply to witness a natural history event of stunning proportions. Such enthusiasts constitute an important constituency with a keen interest in preserving these vital stopover islands. But the fact that the most heavily used areas are quite restricted in space and relatively few in number renders them extremely vulnerable to localized disasters, both man-made and natural. Why are the places where migrating shorebirds concentrate in such numbers so few and far between? The answer to that question and much more are to be found in the pages that follow. —*K.P.A.*

On my first visit to Alaska's Copper River Delta in 1975, few people in Cordova paid more than passing attention to the hordes of shorebirds that annually invaded the mud flats out on the delta or along Orca Inlet, where that small town is perched. Clam diggers and fishermen knew that the arrival of the "snipes" was synonymous with spring [1], but there was little thought about where the birds came from or were going to, much less about their names.

Much has changed in 20 years. Walk down Cordova's main street in early May and stores decorate their windows with silhouettes of sandpipers, banners welcome locals and visitors alike to the annual Copper River Delta Shorebird Workshop, which is sponsored by the chamber of commerce, and buses with volunteer naturalists ferry bird-watchers and the "just curious" out for forays on nearby tidal flats. The birds haven't changed, so what accounts for this heightened level of interest? A bird-loving commercial fisherman named Pete Isleib, the need for environmental-impact studies on oil and gas development projects in Alaska, and ornithology—in that order.

It was the late Pete Isleib, with his contagious enthusiasm for and knowledge of birds, who began to pique the curiosity of his fellow townspeople and tell the ornithological world about the treasures of the Copper River Delta. Pete's stories of pilots flying over the delta and seeing the entire mudflats under them suddenly take wing and turn into clouds of sandpipers caught the attention of government and industry biologists who were tasked with assessing the environmental impacts of oil pipelines and other energy developments in Alaska (Plate 9.1). Finally, with financial support from the sponsors of those energy projects, university professors and graduate students came along to supply the science that firmly put the Copper River Delta on at least the ornithological map.

In late April through about the third week of May, millions of shorebirds use the marshes, sloughs, sandbars, and sedgeflats and mudflats of the Copper River Delta and the adjacent Orca Inlet, which is just around the corner in Prince William Sound. In the spring of 1973, Isleib placed the total number of migrant shorebirds at

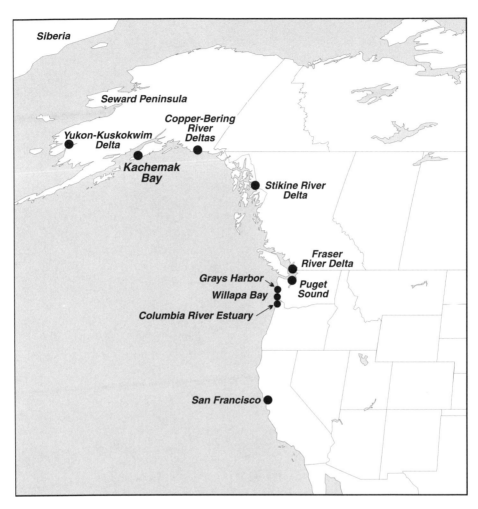

Map 9.1 Notable stopover and staging areas for migrating shorebirds along the western coast of North America.

about 13 million and, though there now is second-guessing about the actual numbers, no one argues that the magnitude is in the millions.

Isleib recorded at least 36 species of shorebirds in the Copper River Delta–Prince William Sound region; and 23 of these were regular, occurring in noticeable volumes [2]. If shorebirds had the equivalent of the route maps found in the backs of airline magazines, their map for Cordova would be a "hub" for northbound flight lines from exotic locations: Pacific golden plovers from Hawaii, bristle-thighed curlews from islands in the South Pacific, surfbirds from Chile, and red-necked phalaropes from the high seas. The main line of flight is the long corridor of the Pacific coast, with feeder lines joining from wetland wintering and stopover sites along the length of two continents. Numerically, the frequent flyers on this coastal route are the western sandpiper and the Pacific race of the dunlin. In combination, these two species easily account for 90% of the migrant shorebirds stopping in the Copper River Delta [3].

In western North America, south of the Copper River Delta, only seven coastal wetlands (Map 9.1) have been identified as supporting concentrations of 100,000 to 1,000,000 shorebirds in spring and fall migration [4]. Spend a few minutes with a relief map of the Western Hemisphere and you can quickly appreciate why. Compared to the Atlantic coast, the Pacific coast of the Americas (i.e., both South and North America) is rather straight, and less fresh water flows into the ocean [5]. The result is that coastal wetlands (estuaries with marshes and tideflats) are few. Most of the shore is rocky or has sandy beaches, which are not nearly so rich in invertebrates as estuarine-wetland habitats. Adjacent uplands typically are forested, and mountains and glaciers rise abruptly, often from the water's edge, especially north of Vancouver. Seen from the air—as by a shorebird in flight—the few significant wetlands along the Pacific coast appear as "islands" of habitat in a coast that is otherwise rather inhospitable, except to the handful of specialists on rocky shores, such as surfbirds and wandering tattlers.

A Project Is Born

On a clear day, the image of the Copper River Delta as an island of habitat is dramatic. On my first flight into the Cordova airport, traveling east from Anchorage, the jet crossed over the 100-mile (about 150 km) width of Prince William Sound with its deep fjords, glaciers, and rocky, forested shores. As we began to descend, we swung in a wide arc over the Gulf of Alaska and then back toward land on what seemed like a different planet, first over sandy barrier islands and then over the vast, more than 400-square-mile expanse (>1,000 sq km) of the low delta of the Copper River. The Chugach Mountains rose abruptly in the background. Streams—spawned from gleaming glaciers and sloughs off the main channel of the Copper River—crisscrossed the snow- and ice-covered, nearly treeless delta.

Perhaps first impressions are always the strongest (I was then at the tender and impressionable age of 24), but I will never forget the birds on that visit in late April and early May 1975. I have never before, nor since, had such a keen sense of the rhythm and spectacle of migration. Although there were new birds to distract me—eerie calls of varied thrushes and rufous hummingbirds whizzing among the Sitka spruce—it was the mass movement of birds that was so captivating. The procession of migrants seemed endless: flock after flock of tundra swans, Canada geese, white-fronted geese, northern pintails, and sandhill cranes calling and heading west and north, day and night. Even better, though, was what was beyond the last of the delta's then frozen marshes, sloughs, and sedge flats. There in the distance were the dense, swirling clouds of sandpipers, just as Isleib had described them. These were truly the "wind birds," so aptly named by Peter Matthiessen [6].

I was there with David Norton to check out prospects for my graduate thesis project on shorebirds for the University of Alaska at Fairbanks. Other than Isleib's accounts of millions of sandpipers, almost nothing was known about shorebirds' use of the delta. In fact, in the mid-1970s, knowledge of the biology and ecology of many Western Hemisphere shorebirds was very limited—largely based on a handful of studies on breeding grounds in Alaska and Canada. Migration studies on shorebirds were scarce indeed [7].

The impetus for my shorebird project-in-the-making was that the federal government was proposing to sell oil and gas leases on the continental shelf ringing much of Alaska, including that in the northern Gulf of Alaska. In addition, oil from Alaska's North Slope would soon start flowing through the Trans-Alaska Pipeline and be transported by tanker through Prince William Sound into the northern Gulf of Alaska, immediately west of the Copper River Delta. The waters of the north gulf circulate in a large counterclockwise gyre, so there was concern that oil spilled in the gulf—whether from a tanker or a drilling rig—might sweep across the delta from east to west. Through such programs as the Federal Bureau of Land Management's Outer Continental Shelf Environmental Assessment Program, funds were then available for graduate students and others to explore these concerns in advance of the development decisions.

Because so little was known about shorebird migration in general, and about migration in the Copper River Delta specifically, we came to Cordova asking simple questions: Why are millions of shorebirds stopping here? What are they eating? How do they use the area? How important is the area in relation to others? What strategies do shorebirds use in their migration, and how does the Copper River Delta fit into these strategies? What would happen if the delta's habitats were not available, owing to alteration or pollution? As a graduate student, I was fortunate to be able to address questions that were of ornithological interest and that were timely and important from the standpoint of conservation and management.

When my studies began in earnest in 1976, the basic plan was simple in concept. Long strip-transects were to be marked with stakes at representative sites on the mudflats. These transects then provided a basis for a) censusing the numbers and types of shorebirds, b) watching their behavior at different tide stages, c) collecting a small number of specimens to determine what they were eating and how fat they were, and d) sampling the mudflats themselves to see what potential prey they contained. While my intensive work focused on these transects, I also chartered time in small planes to gauge the numbers and distribution of birds across the vast Copper River Delta, as well as in the neighboring Bering River Delta a few kilometers to the east.

It is easy now to write that my basic study plan was simple in concept, but I well remember how difficult it was to carry out. Most of the delta and its extensive mudflats are inaccessible by road, so one gets around on foot or by boat or light plane, each of which has its own hazards. By foot, the distances are great, the mud is often slick, and the tideflats are crisscrossed with sloughs that are deceptively deep and treacherous. The tides can range more than 20 feet (about 6 m), and knowing exactly what the tide is doing is crucial to personal safety, to say nothing of understanding shorebird behavior and ecology.

Fortunately, I was able to drive to a key study site at Hartney Bay, which is today the prime viewing location for shorebirds near Cordova. Even at this site, however, the water's edge at low tide was a mile or more from the road. One day a clam digger became disoriented by the sudden descent of a dense fog and was drowned in one of the sloughs when the tide came in. I was there—in fact, probably not far away—but I heard not even a cry for help.

The logistics of movement about the delta are complicated by the weather. Annual precipitation averages about 81 inches (206 cm); much of it comes horizontally, driven by the wind. Temperatures during the spring shorebird migration tend to hover near freezing, with storms sweeping in off the Gulf of Alaska every few days. Eyeglasses, binoculars, and scopes fog up, hip boots or "Cordova sneakers" (knee-high rubber boots) are usually wet, and rain gear becomes an exoskeleton that you long to shed. None of this seems to bother the birds very much, although I have seen western sandpipers seeking shelter underneath and between the legs of black-bellied plovers and whimbrels—and those same large shorebirds tumbling head over heels, driven by the wind. Although the logistics were difficult and the weather a challenge, it didn't take long to develop at least a basic picture of what migrating shorebirds are doing in the Copper River Delta in spring.

Local breeders—greater and lesser yellowlegs, common snipe, short-billed dowitchers, and least sandpipers—spend little time on the tideflats. For the passing throngs of migrants, however, most of the springtime action is on the tideflats. In some years, in fact, snow and ice cover much of the wetlands above the high-tide line when the shorebirds first arrive, and tideflats are all that is available until well into May.

They Come to Eat

If one flies from Kayak Island in the east (where in 1741 George Steller was the first European naturalist to set foot in Alaska) to Cordova in the west, across the lengths of the Bering and Copper river deltas, it quickly becomes evident that not all of these vast wetland complexes are used heavily by migrant shorebirds. The birds largely steer clear of what superficially appears to be extensive good habitat around the outlets of the main river channels. The tidal flats at these sites are scoured by water- and wind-borne sediments and by ice coming downriver from the interior. Such areas must be inhospitable to birds; but, even more important, they must be inhospitable to what birds eat. And therein lies the key to what brings shorebirds to the Copper River Delta (Plate 9.2). They come to eat!

The more sheltered flats of the Copper River Delta are rich in intertidal life. A walk across Hartney Bay at the midtide level is convincing. The surface is littered with clam shells; live mussels and barnacles are attached to small rocks and flotsam; the holes from the siphons of live clams and the tiny conical mounds deposited by polychaetes (marine worms) are myriad; and amphipods (small crustaceans) scoot across thin slicks of water left on the mud by the receding tide. The tideflats literally crunch as you walk, and tiny jets of water, expelled through the siphons of live clams, spout ahead of you.

In terms of total biomass (i.e., the combined weight of all of the individual organisms), the most important organism is a tiny clam, *Macoma balthica*. Our observations of feeding birds and examination of the stomachs of sandpipers collected in the Copper River Delta quickly revealed that *Macoma* was the preferred prey for dunlins and very important for western sandpipers. Indeed, *Macoma* is found in northern estuaries around the world and is a major source of food for dunlins and other shorebirds at such places as the United Kingdom and the Netherlands [8]. A good-sized *Macoma* is about as big as the nail on one of your little fingers; a giant *Macoma* would be about as wide as a thumbnail. They are plucked out of the mud by the probing beaks of shorebirds and swallowed whole. The bird's gizzard breaks and grinds the shell, which is later regurgitated — "cast out" — as a pellet. The clams' substantial and nutritious soft parts are then digested by the bird.

Our research revealed that dunlins in the Copper River Delta preyed heavily, perhaps almost exclusively, on small bivalves, especially *Macoma* [9]. The Pacific dunlin has a rather long beak, up to about 1.5 inches (about 40 mm), which is ideal for reaching the clams in their burrows. The western sandpiper has a shorter beak, at most a bit more than an inch (about 30 mm), and we found that this species took a wider variety of prey, including many *Macoma* but also many amphipods, larval insects, snails, and even copepods. This difference in diet was reflected in the behaviors of these two species on the tideflats.

On rising and high tides, western sandpipers and dunlins both roosted in large, mixed-species flocks on the tideflats. If conditions were calm, these flocks would be out on the open flats, often near the water's edge, moving as needed to stay above the advancing tide. Under windy conditions, the shorebirds were strung out beneath the top plane of the sloughs that crossed the tideflats, or they sought shelter in the lee of logs or some shoreline feature. Within a roosting flock, dunlins would strike classic poses of birds at rest—head laid back along a wing, often standing on one leg—or they would preen. On the other hand, many western sandpipers fed continuously through a high-tide period; others would punctuate their rest with bouts of feeding, picking at whatever was nearby. This different behavior began to explain the more diverse diet of the western sandpiper compared to the dunlin.

Once the tide started to go out, more and more sandpipers began peeling away, although it typically took about an hour after the tide had begun to recede before all birds resumed feeding. During the falling tide, the pace of feeding was urgent; the birds were all business. Dunlins almost invariably crowded about the water's edge—the line of the receding tide—probably because they could cue in on the still-extended *Macoma* siphons, which showed the dunlins where they could probe most profitably. The dunlins were so intense in their feeding that they aggressively crowded out the smaller western sandpipers, sometimes displacing them from the zone immediately about the tide line. As a result, western sandpipers would be found scattered at various elevations across the now exposed tideflats, feeding on a wide variety of prey.

At low tides, the shorebirds dispersed widely, and often seemed to disappear altogether for a time. Fewer and fewer dunlins fed as the tide began to rise again; western sandpipers continued to feed regardless of the tide stage. Why did western sandpipers feed almost constantly on a wide variety of prey, while dunlins fed most intensely during the falling tide and took primarily a single species of prey? Although both birds are small, a dunlin is typically about twice as heavy as a western sandpiper. Apparently, its long beak gave it access to the abundant supply of *Macoma* beneath the mud's surface and enabled it to satisfy its energy needs with what it obtained during a few hours of intense feeding. The smaller western sandpiper, which burns its fuel more quickly than the dunlin (smaller birds lose heat faster and have higher metabolic rates per unit of mass than larger birds) and which does not have such ready access to the *Macoma*, needed to feed more or less continuously to sustain itself.

Energetics is the key to understanding the significance of the Copper River Delta and other such stopover areas in the annual cycles of shorebirds like western sandpipers and dunlins. Recall that the Pacific coast has few large wetland complexes, and the wetlands it does have are widely separated islands of habitat strung along thousands of kilometers of shoreline.

Fat is the fuel for long-distance migration in birds, whether they are ruby-throated hummingbirds crossing the Gulf of Mexico, blackpoll warblers crossing the Atlantic Ocean between the Maine coast and South America (see Chapter 5), or dunlins flying nonstop from British Columbia's Fraser River Delta to Alaska's Copper River Delta. Birds undertaking long flights must have substantial fat reserves, and the longer the nonstop flight the more depleted are the reserves upon arrival. Wind conditions, whether calm, adverse, or favorable, play a crucial role in how much energy is required en route [10].

In my initial studies, an examination of the weights of northbound dunlin and western sandpiper specimens in museum collections revealed that dunlins, especially, were highly variable; but they were very heavy, with large fat reserves, at such locations as the Fraser River Delta in British Columbia. These data led me to suggest that many if not most dunlins probably flew nonstop to the Copper River Delta. I found less dramatic extremes in the weights of western sandpipers (i.e., fewer very fat and very lean individuals), and I suggested that this smaller species perhaps undertook shorter flights than the dunlin, stopping more frequently along the way [11]. In any case, both species were dependent on a limited number of coastal stopover sites to fuel their migrations north.

Because of the tremendous numbers of western sandpipers and dunlins stopping in the Copper River Delta, we suspected that virtually the entire world populations of western sandpipers and the Pacific race of the dunlin were obliged to stop there each spring. Migrants would reach the delta by flights of varying distances: for example, a direct flight from the Fraser River Delta would be about 1,250 miles (about 1,900 km). Regardless of the length of the nonstop flight before arrival in the delta, however, all migrants would presumably arrive with fat reserves that were depleted to some degree. These reserves then would be restored through voracious feeding on *Macoma* and the other rich invertebrate life of the tideflats, as described above.

Isleib had suggested that the "turnover" in migrating shorebirds in the Copper River Delta was rapid, and knowing how long an individual or a flock remained in the delta was key to understanding how many birds used the area and the importance of the energy they obtained. With my colleagues David Norton and George West, I approached this question by glueing small colored feathers to the tails of western sandpipers caught live in mist nets (Plate 9.3). We chose this technique because radio transmitters were then too large to place on a bird the size of a western sandpiper, which weighs only about 28 g (about 1 oz). Although the technique was primitive, particularly when rain made it hard to see a single soggy feather attached to a distant flying bird, we established that western sandpipers remained at such sites as Hartney Bay for 2 to 6 days before continuing their migrations [12]. We believed that these results were probably maximum lengths of stay.

To review briefly, the basic picture that emerged from my research in the mid- to late 1970s, was that the Copper River Delta was a critical stopover site for huge numbers and entire populations of western sandpipers and dunlins on the Pacific coast. Dunlins reached the delta by extended, nonstop flights, whereas western sandpipers probably stopped more frequently en route. Both species stopped in the delta for several days, using rich intertidal feeding opportunities to replenish depleted fat reserves. Dunlins fed intensely during falling tides, preying almost exclusively on *Macoma*, whereas western sandpipers fed throughout the tide cycle and ate a variety of invertebrate prey.

Birds on a Mission

This basic picture has remained intact over the past two decades; fortunately, too, others have picked up where this early work left off. Thanks to more recent and ongoing studies by Mary Anne Bishop, Rob Butler, Robert Gill, George Iverson, Nils and Sarah Warnock, and others, understanding of shorebird migration, especially that of western sandpipers, has greatly improved. An important breakthrough has been the development of radio transmitters weighing only about one gram, which means that they can be carried safely by western sandpipers.

In 1992 (and, more recently, again in 1995 and 1996), Iverson and his colleagues affixed radio transmitters to dozens of western sandpipers at several sites on the Pacific coast, including at San Francisco Bay, California [13] (Plate 9.4). With the help of cooperators armed with radio receivers at sites spanning a 4,000-kilometer stretch of coastline, they then followed the sandpipers' progress north. Fifty-eight of the original 77 western sandpipers fitted with radios in 1992 were relocated en route. About 62% of those 58 were relocated again in the Copper River Delta, thus confirming its significance for this species. Several birds that stopped in the Fraser River Delta or Stikine River Delta, however, were not detected in the Copper River Delta, indicating that at least some western sandpipers can overfly this stop.

Following the radio-tagged birds also confirmed that most western sandpipers employ a short-hop strategy, stopping for an average of three days at each of several sites en route north. For the western sandpipers tagged at San Francisco Bay, the trip to the Copper River Delta took about 12 days, suggesting 3 or 4 stops along the way. One individual made that same flight in less than 42 hours (1,850 km/d; 77 km/h), so some western sandpipers do undertake extended nonstop flights!

Iverson and his colleagues [13] concluded that the western sandpiper's "short-flight strategy emphasizes the importance of maintaining a continuous complex of intertidal wetland habitats along the migration route to ensure shorebird conservation." The Copper River Delta is one such habitat area along the Pacific coast migration route. But, since the delta is a stopover—not a final destination—for western sandpipers and dunlins that are northbound on the long Pacific corridor, this is not quite the whole story.

If you are fortunate enough to have watched large numbers of shorebirds in migration, you may have seen the incredible gyrations of their flocks, wheeling over the tideflats in synchrony, at breakneck speeds. Shorebirds feeding on the mudflats will at times, with no visible warning, leap into the air in unison and whirl spectacularly over the flats like roller coasters. In my experience from the Copper River Delta, "these outbursts seem to come most often on rising tides, especially in the evening hours, when good feeding areas are covered. After several wide swings about the flats, they may suddenly spiral upward, all twittering excitedly, and once again head west and north" [14]. The sense of urgency in these flights is unmistakable. These are birds on a mission.

The dunlins and some of the western sandpipers leaving the Copper River Delta are sufficiently fat that they can fly nonstop another 800 to 1,000 km to breeding grounds in western Alaska, such as the Yukon-Kuskokwim Delta, which is where most western sandpipers and Pacific dunlins nest [15, 16]. Others may reach the Seward Peninsula and some western sandpipers will cross the Bering Strait to Siberia. Consistent with their short-hop strategy, however, many western sandpipers make an additional stop in lower Cook Inlet, at places like Kachemak Bay, before undertaking the final leg of their journey north.

The schedule is tight. The first western sandpipers and dunlins, usually males, arrive on the tundra by 10 May, plus or minus a few days. The males set up and defend territories and compete for the attention of females as they arrive. Courtship must be completed, nests initiated, and clutches of eggs complete in the first two weeks of June. There are few second chances at these northern latitudes.

The energy demand on the breeding adults is enormous, but under good conditions there will be at least some spiders and early insects to sustain a dunlin or western sandpiper when it first arrives on the tundra. If snow and ice still prevail, however, or if there is a late storm, what is there to enable the new arrivals to survive, much less to initiate breeding? The answer, of course, is fat. The same fuel that carried the tiny sandpipers north should carry them through the first few days on the tundra, before hordes of insects start to emerge.

Some of the individual western sandpipers radio-tagged by Bishop and Warnock have been tracked all the way to their breeding grounds on the Yukon Delta! So far, though, there has been no chance to follow the fates of individually radio-tagged breeders, nor to assess their success relative to their body condition at the start of migration or at stopovers along the way. Research by Declan Troy, on shorebirds nesting on the North Slope of Alaska, shows a) how the proportion of birds not even attempting to breed in a given season is a function of the weather and b) how the loss of a single year's worth of productivity can mean a smaller breeding population two breeding seasons hence [17]. So, even while more research is always needed and indeed should be undertaken, the results of all these studies to date make the importance of the Copper River Delta and each of the other wetlands remaining on the Pacific coast abundantly clear. It is a simple, no-rocket-science equation: *To a migratory shorebird, a wetland = food and rest. The food is converted to fat, and the fat = fuel for migration and energy to sustain it during the early breeding period.* Under good conditions, sandpipers arriving on the breeding grounds may not need large deposits of fat. Under poor conditions, which are by no means infrequent, the amount of fat may make the critical difference.

The Western Hemisphere Shorebird Reserve Network, conceived and initially organized by John Peterson Myers, is one means by which the importance of the Copper River Delta and other stopover sites has been recognized [18]. The network, however, has no special protective authority, and pressures on coastal resources and habitats are enormous and growing [19]. In Alaska, it is the never-ending human quest for energy that led to the shorebird studies in the Copper River Delta and that continues to pose the greatest threat. Although oil and gas leasing on the continental shelf offshore and east of the delta are not imminent, the possibility remains, and the delta is always at risk from a tanker accident. Shorebird staging areas in nearby lower Cook Inlet are at risk from both offshore drilling and the passage of tankers.

The oil spilled by the supertanker *Exxon Valdez* did not reach the Copper River Delta, but there was a near miss: tens of thousands of surfbirds and black turnstones (Plate 9.5) were staging on Montague Island, only a few kilometers west of the delta and barely out of the oil's path in western Prince William Sound [20]. For me, this event more than any other is a reminder of how man's mistakes can quickly turn into nightmares for the birds. For now, the passage of western sandpipers and dunlins across the Copper River Delta is part of a parade that spans the long coastline of two continents, reaching even a third. With luck and, more important, with care, the arrival of the "snipes" will continue to be synonymous with spring in Cordova.

References

1. Isleib, M.E. 1979. Migratory shorebird populations on the Copper River Delta and eastern Prince William Sound, Alaska. Stud. Avian Biol. 2:125–129.

The initial study that drew attention to the vast concentrations of shorebirds at the Copper River Delta.

2. Isleib, M.E., and B. Kessel. 1973. Birds of the North Gulf Coast–Prince William Sound Region, Alaska. Biol. Papers Univ. Alaska, No. 14. Fairbanks.

A general distributional survey of the birds of the region.

3. Senner, S.E. 1977. The Ecology of Western Sandpipers and Dunlins during Spring Migration through the Copper Bering River Delta System, Alaska [M.S. thesis]. Fairbanks: University of Alaska.

4. Page, G.W., and R.E. Gill. 1994. Shorebirds in western North America: late 1800s to late 1900s. Stud. Avian Biol. 15:147–160.

A readable examination of a century's worth of change in shorebird populations.

5. Pitelka, F.A. 1979. Introduction: the Pacific coast shorebird scene. Stud. Avian Biol. 2:1–11.

A general overview by a pioneer in the study of shorebirds along the western coast of North America.

6. Matthiessen, P. 1973. The Wind Birds. New York: Viking Pr.

An exquisite essay on shorebirds by a master writer on natural history subjects.

7. Recher, H.F. 1966. Some aspects of the ecology of migrant shorebirds. Ecology 47:393–407.

A technical paper dealing mainly with feeding and territorial behavior during migration.

8. Green, J. 1968. The Biology of Estuarine Animals. Seattle: Univ. Washington Pr.

A general text.

9. Senner, S.E., D.W. Norton, and G.C. West. 1989. Feeding ecology of western sandpipers, *Calidris mauri,* and dunlins, *C. alpina,* during spring migration at Hartney Bay, Alaska. Can. Field-Naturalist 103:372–379.

Technical data.

10. Butler, R.W., T.D. Williams, N. Warnock, and M.A. Bishop. 1997. Wind assistance: a requirement for bird migration? Auk 114:456–466.

A technical consideration of the importance of wind in the success of migratory flights.

11. Senner, S.E. 1979. An evaluation of the Copper River Delta as a critical habitat for migrating shorebirds. Stud. Avian Biol. 2:131–145.

An environmental assessment of the importance of the Copper River Delta for passage migrants.

12. Senner, S.E., G.C. West, and D.W. Norton. 1981. The spring migration of western sandpipers and dunlins in southcentral Alaska: numbers, timing, and sex ratios. J. Field Ornithol. 52:271–284.

Technical data.

13. Iverson, G.C., S.E. Warnock, R.W. Butler, M.A. Bishop, and N. Warnock. 1996. Spring migration of western sandpipers along the Pacific coast of North America: a telemetry study. Condor 98:10–21.

A technical paper describing the results from tracking radio-tagged western sandpipers.

14. Senner, S.E. 1979. Delta of the wind birds. Living Wilderness 42:22–27.

An essay on the Copper River Delta and its shorebirds; written for a general audience.

15. Wilson, W.H. 1994. Western sandpiper (*Calidris mauri*). In The Birds of North America,

No. 90 (A. Poole and F. Gill, eds.). Philadelphia and Washington, DC: Acad. Nat. Sci. and Am. Ornithol. Union.

16. Warnock, N.D., and R.E. Gill. 1996. Dunlin (*Calidris alpina*). In the Birds of North America, No. 203 (A. Poole and F. Gill, eds.). Philadelphia and Washington, DC: Acad. Nat. Sci. and Am. Ornithol. Union.

    This and the preceding reference are the most up-to-date life histories of these two species.

17. Troy, D. 1996. Population dynamics of breeding shorebirds in Arctic Alaska. Int. Wader Stud. 8:15–27.

    A technical paper.

18. Myers, J.P., R.I.G. Morrison, P.Z. Antas, B.A. Harrington, T.E. Lovejoy, M. Sallaberry, S.E. Senner, and A. Tarak. 1987. Conservation strategies for migratory species. Am. Sci. 75:19–26.

    An excellent overview of conservation issues relevant to migratory birds, with a special emphasis on shorebirds.

19. Senner, S.E., and M.A. Howe. 1984. Conservation of Nearctic shorebirds. Behav. Mar. Organisms 5:379–421.

    A readable discussion of conservation issues involving shorebirds.

20. Norton, D.W., S.E. Senner, R.E. Gill Jr., P.D. Martin, J.M. Wright, and A.K. Fukuyama. 1990. Shorebirds and herring roe in Prince William Sound, Alaska. Am. Birds 44:367–371, 508.

    A readable account of the birds and one source of food.

10 WILLIAM A. CALDER

Hummingbirds in Rocky Mountain Meadows

Banding birds, especially small migratory birds with short life spans, is a real act of faith. The chances of any one of these embellished creatures ever having another close encounter with a human being are vanishingly small; and yet the apparently inconceivable does happen, as Jim Baird recounted in Chapter 5. Still, such occurrences are the rarest of rare events. The prospect of demonstrating site fidelity during brief migration stopovers seems even more remote for hummingbirds refueling in alpine meadows. They are very tiny birds that live very short lives, and there are many flower-covered mountain meadows that would seem to provide suitable habitat. But in addition to learning much about the behavior and ecology of rufous hummingbirds during migration, Bill Calder has shown conclusively that these "red menaces" are able to navigate with sufficient precision that at least some of them can visit the same patch of flowers on subsequent fall migrations. This is but another humbling reminder that there remains much about bird migration that we cannot fully explain.

Hummingbirds are fundamentally a tropical group. Few species have managed to penetrate far into higher latitudes. To do so requires that they evolve seasonal migration. Interestingly, several tropical species have also been found to engage in regular migrations, though none goes so far as the ruby-throated hummingbird, which crosses the Gulf of Mexico in a single flight, or the rufous hummingbird, the species with the longest migration. Place yourself now in one of those lovely high valleys filled with wildflowers, and watch through Bill Calder's eyes the remarkable migration of North America's smallest long-distance traveler. —K.P.A.

The Setting

In the steep terrain of southwestern Colorado, a river can lose its identity not far from its origin. Series of mergers with adjacent rivers are not unlike corporate mergers, leading to a larger corporation with a different name. The East River drains everything inside the facing ridges, which connect seven glacier-sculpted peaks whose summits top 12,000 feet (3,600 m). In a mere 25 miles (40 km) the river runs down a steep-walled valley, meanders when the widening valley bottom affords that luxury, and then becomes a subsidiary of the Slate River. The Slate flows only 8 miles (13 km) farther before a merger with the Taylor to become the Gunnison River, which flows 146 miles (225 km) to join the Colorado.

Like the cold water, cold air also flows down these channels. In the winter, Gunnison, the first official weather station in this cold-air drainage is often "the coldest spot in the nation" on the morning news—definitely not a suitable winter home for hummingbirds! But come back half a year later and we'll stand in a meadow of flowers clothing the glacial till above the East River (Fig. 10.1). The sunny, mid-July morning does not remember the ice that covered it in the Pleistocene; it has even forgotten the chill of winter weather reports! Aspen clones cover nearby slopes, except where avalanches have blown them down or bulldozed through them. A meadow stage has persisted down in the 1878–1910 mining town of Gothic (9,540 ft, or 2,910 m), afforestation having been suppressed by some combination of human activity, cattle grazing, winter cold, snow patterns, and rodent foraging. Gothic has been the Rocky Mountain Biological Laboratory (RMBC) since 1928.

If there is one word to describe summer at high elevations, it is *intense*. Summer cannot play for five months here, as it does east of the Rockies. Short growing seasons cram brilliant flower displays into a brief "window of opportunity" between the late spring melt and the first killing frost. Constrained by short-term flower abundance, timing is crucial for the nectar-feeders: insects and our three species of hummingbirds.

Of 17 hummingbird species that have bred in the United States, the rufous is the

Figure 10.1 A flower display in the Colorado Rockies awaiting pollination by transient humming-birds. (Photo by W. A. Calder)

best choice for the study of hummingbird migration. In the wide belt of more than 900 miles (>1,500 km) separating the southern limit of breeding from the northern limit of the wintering range, any rufous observed is clearly a migrant.

The rufous is the most studied hummingbird, especially when it comes to its physiology, thanks to Carol Beuchat, Terry Bucher and Mark Chappell, Lynn Carpenter, Lee Gass, Sara Hiebert, and Raúl Suárez, to name a few. Large numbers of southbound rufous, refueling on the seasonal abundance of subalpine and alpine flowers in July and August, have made them a model system for investigating foraging behavior and pollen transfer. Thus there is more background information to help us interpret what we observe in the field [see Refs. 1 and 2 for a general introduction to hummingbirds and their life-styles].

Rufous and broad-tailed hummingbirds probably have a shared ancestry in the central highlands of Mexico. As the Pleistocene glaciation retreated, more northern hummingbird habitat was created by plant succession. "Broad-tails-to-be" extended their range up the Rockies to Utah and the Greater Yellowstone ecosystem. They now inhabit the highest elevations from western Wyoming south to Guatemala. "Rufous-to-be" dispersed up through the Pacific coastal states, colonizing Oregon and Washington seasonally, spreading inland to Idaho, western Montana, interior British Columbia, Alberta, and northward. Eventually, they claimed coastal Alaska to Prince William Sound, becoming the hummingbirds of highest latitudes (Plate 10.1). Between Alaska and the winter range in central and western Mexico, the species has the distinction of flying the longest migration of any hummingbird, and in terms of body lengths (almost 49 million), the longest migration of any bird [3, 4].

The zone between breeding and winter distributions of the rufous hummingbird does not remain vacant habitat until the rufous migration. Broad-tails have colonized the middle and southern Rockies and range through Nevada and eastern California. Body masses, bill lengths, and tongue reaches of broad-tails and rufous are very similar, so their demands for the same food resources are likely to overlap (Plate 10.2). When rufous transients come through, competition and interspecific conflict are inevitable. Natural selection seems to have made rufous very aggressive. Their primary goal, to fatten and fly, puts great demand on the flowers upon which local breeding broad-tails rely. Neither species can be appreciated without quantitative knowledge of the other.

How do their numbers compare? In 12 years, we captured an average of 229 broad-tailed and 345 rufous hummingbirds per year. The rufous passage lasted about 6 weeks and individuals stayed 1 to 2 weeks to refatten.

The broad-tails have Gothic to themselves from mid-May until July, when the transient rufous show up. Initially, broad-tails come up daily from the security of a more reliable energy base camp, lower in the valley, to check on the progress of snowmelt and flower emergence around Gothic. Usually before May ends, a successful male will find a territory that provides enough energy to make ends meet, so he stays and makes his claim. Territorial defense and courtship occupy the male's days for the next six weeks. Females move up simultaneously or lag slightly behind the return of early males. By June's end, most females are finishing the 16-day to 19-day incubation of their two eggs; some are brooding and feeding chicks. Proteins and salts missing from a wholly nectar diet are provided by swarms of dancing-flies and gnats on the wing and herds of aphids on the vegetation.

Those flowers that matter to local hummingbirds flash by in a rapid blooming sequence. We associate hummingbirds with red flowers, but these hummingbirds associate blue-purple larkspurs with nectar one-third richer in sugar than local red flowers. Such super-richness is timed perfectly, being available when energy demands are greatest: early in the season, when nights are coldest. Nelson's larkspur blooms. As these fade, red-flowered plants await hummingbird pollination and provide nectar for midseason. Red columbine are found in moist, shady sites along stream banks and on the north sides of moraines left by valley glaciers. Twinflower honeysuckle grows in moist but sunny situations. The broad-tails just fight among

themselves for these three flower species, since rufous are not in town until blooming ceases. Thriving a bit later, in open, drier habitats, are patches of scarlet gilia and
Indian paintbrush. In July's energy crunch when hummingbirds battle over nectar
supplies, the other purple larkspur, Barbey's, blooms and offers 45% sucrose nectar.

In early July, the "red menace" descends! Male rufous hummingbirds, with backs
rusty-red like a robin's breast, invade the scene and probe the flowers, on average
one week ahead of the females and two weeks ahead of the juveniles. The adults had
reached their northern sites earlier, bred earlier, and now head southward to Mexico
earlier than the broad-tails. The rufous have a special urgency for reaching their winter range. Land area decreases along their journey as North America tapers southward to an isthmus; species diversity increases with this drop in latitude and space.
With more hummingbirds of more species overwintering in less area, it pays to be
early and aggressively territorial.

Without the interloping rufous, resources seem adequate for female broad-tails
to complete the nesting season with ease. But when refueling in Colorado, the rufous draw on the same resources that broad-tailed mothers count on for finishing
their single-parenting. With rufous transients draining the flowers, some female
broad-tails, unable to meet the needs of their nestlings, are forced to abandon them.
The parent's only hope is to survive and try again after another round-trip to Mexico. Chasing more but enjoying it less, male broad-tails are ready to give up territorial claims until another year [5].

American robins, white-crowned sparrows, and yellow warblers were in full song
just a month ago, but now the bickering of rufous hummingbirds fills in the vacated
airwaves. With a timbre suggesting a miniature eastern kingbird, an irritated "eee-
didjer didjer didjer" emanates from the larkspur clump to wave off another hummingbird on final approach for a meal. We scan the meadow to find hummingbirds.
They are perched atop or darting between clumps and are hovering at flowers all
over the place—there must be dozens upon dozens of them. Not all territorial birds
are the rusty-red adult males that first caught our attention. Green-backed females
and even juveniles can be almost as assertive when their survival is at stake.

Over in Washington Gulch, which drains Gothic Mountain's western flank, is the
Painter Boy mine, an eyesore heap of tailings topped by a small, collapsing mill
building now occupied by a fat yellow-bellied marmot. A rusting old International
pickup sits outside, propped up, without its wheels, where it once served as the
mine's power plant. A small stream trickles past the base of the heap and disappears
under an adjoining lush cover of soft, fleshy *Corydalis*, relative of the familiar bleed-
ing heart and Dutchman's-breeches (Plate 10.3).

Hummingbirds like *Corydalis*, even though its dull ivory-white flowers with pale
purple- pink edging cannot match the brilliance of scarlet gilia and paintbrush or

deep-purple larkspur. What really matters is the nectar. My samples were 32% to 34.5% sugar (by mass, g/g), comparing favorably with most other local nectars. Nick Waser [personal communication] found a much higher average sugar concentration of 50.2% (better than larkspurs!) in *Corydalis* nectar in Washington Gulch in 1984. No wonder the rufous hummingbirds are fighting over patches of this strange plant! If rufous migrants pollinate *Corydalis* effectively, they also deserve credit for protecting a recovering watershed and its water quality.

Calliope hummingbirds (Fig. 10.2) usually pass through the meadows simultaneously, but are scarcely noticed because they are less assertive and tend to perch low in the vegetation during migration. Calliopes and rufous overlap in their breeding range in inland mountains from British Columbia to Oregon and Montana. Calliopes, distinguished when perched by wingtips that extend beyond their stubby tails, are the smallest long-distance avian migrants. Some apparently travel 2,200 miles (3,600 km) one-way, despite their constrained flight speed, fuel endurance, and brain size.

The avian food web at RMBL also includes orange-crowned warblers, which the

Figure 10.2 The much less aggressive and smaller calliope hummingbird shares migration stopover sites with rufous hummingbirds. (Photo by T. Dodson for Cornell Laboratory of Ornithology)

rufous encounter repeatedly during the annual migration circuit. Orange-crowns feed on nectar as well as insects: in July we netted almost as many orange-crowns as rufous hummingbirds in patches of *Corydalis* at the Painter Boy mine. Resource competition is not limited to within the "hummingbird community." At Gothic, rufous and orange-crowns are also among the several species of animals that Paul Ehrlich and Gretchen Daily found competing for willow sap in wells made by red-naped sapsuckers [6]. In winter, hummingbirds and orange-crowns compete for the yellow heads of a large bush-like *Senecio* (ragwort) on Volcán Nevado in Jalisco, Mexico; and in spring migration they compete for chuparosa nectar in Sonora, the orange-crowns tearing apart the blooms to get at the nectar.

The Migration

Watching the migrant rufous feeding and fighting in the meadows of the East River, one wonders where their flights originated, what routes they use, and where in Mexico they will go. Small size is an advantage for survival on nectar, but a disadvantage for seasonal home-changing. Small size puts limits on fat storage and flight speed, and that reduces endurance and range between refuelings. A Canada goose, for example, can fly six times as far between fueling stops, at two or three times the hummingbird's 25-mph speed (no tail wind), so the hummingbird's migration trips take up more of its lifetime. If the fuel supply runs short, the weather changes, or the migrating hummingbird is blown off course or forced down unexpectedly, the unlucky bird might be far from a flowering meadow, making it difficult or impossible to resume on new fat.

Geese have advantages: the pooled experience and route knowledge of flock mates and relatives and the reduction of cost by flying in a V-formation. Because rufous hummingbirds are decidedly antisocial, there is no sharing of route information or energy costs, and their smaller brain might limit storage capacity for landmarks and navigational computations. The goose's comparatively huge brain has about three times the mass of an entire hummingbird, or 66 times the mass of the hummingbird's brain. Smaller brains are more densely packed with nerve-cell bodies. Stanley Cobb showed that the brain of the broad-tailed hummingbird, for example, had almost three times the neuron density of an emu's brain [7]. So, if the goose has a megabyte of migratory-route storage capacity, then the hummingbird, adjusted for size, would have only about 34 kilobytes. Would that be enough for a rufous hummingbird to remember the way from Alaska to Mexico? Wouldn't it be fascinating to know that route, begun only by instinct in a juvenile, and to know the criteria for the best route? What geographical information do hummingbirds know and use for navigation?

Hummingbirds are rarely seen in mid-migration flight. From vantage points at the southern ends of ridges in the Appalachian and Allegheny mountains, and atop a 150-foot-high cliff on the north shore of Lake Erie, migrating ruby-throated hummingbirds have been observed in southward flight by Willimont and colleagues [8]. Fred Sharp has observed rufous hummingbirds departing Cape Flattery, Washington, and then flying low over the Strait of Juan de Fuca toward Vancouver Island, British Columbia, in the springtime [pers. comm.]. Similar flights are made on the return migration. On the ferry on 20 July 1996, about halfway across the straits from Vancouver Island, Cam and Joy Finlay observed two hummingbirds, one a rufous for sure, "flying at eye level (no more than fifty feet off the water) heading straight south to the Washington coast . . . around 7 A.M." Others in the group saw four more rufous moving south [C. Finlay, pers. comm.]. In California, Lynn Carpenter, David Paton, and Mark Hixon watched fattened rufous hummingbirds take off in a "characteristic flight path upon departure, generally high and due south for as long as we could hold them in binocular view. These flights generally began between 0600 and 0800 hr." [9]. The rarity of such observations shows how limited we are if we learn about hummingbird migration only from seeing migratory flight in progress! Hummingbird migrations are known only by a combination of anecdotes, distributional accounts, and what we can observe, count, and capture when they stop to rest and refuel.

Alan Phillips's analysis of collection dates of museum specimens revealed an elliptical ("racetrack") migration pattern for rufous hummingbirds [10]. From the northern Rockies they move down the central and southern Rockies to the Sierra Madre Oriental of eastern Mexico (Map 10.1). In midwinter they move westward along the transvolcanic belt. All rufous seem to return northward along the Sierra Madre Occidental and up the Pacific coastal lowlands and foothills, either a) to continue coastally to British Columbia and Alaska or b) to fly inland to Idaho, Montana, and Alberta.

This annual elliptical route is consistent with patterns in weather and floral phenology. Coastal environments are moderated by the maritime climate. Winters are less severe, and spring comes earlier than it does in the Rocky Mountains. Consequently, rufous can migrate north earlier than broad-tails, and in Alaska they can be nesting a month earlier than a broad-tail in Colorado. This schedule gives them a second advantage: they finish breeding earlier so they can fly south along the Rockies when the summer flowers are in their prime during the monsoons, which send in abundant moisture from the Gulf of Mexico.

Rufous are notorious for fighting over flowers and feeders during the "fall" migration (mostly four to six weeks before the autumnal equinox). The two major "flyways" pass southward on either side of the Great Basin Desert: 1) along the Coast

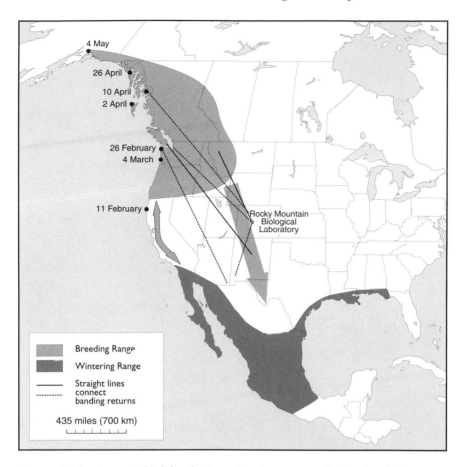

Map 10.1 Rufous hummingbird distribution, migration routes, and migration timing. Straight lines connect banding and recovery sites. Solid lines pertain to birds captured in the same year that they were banded; dashed lines pertain to birds captured in some later year, and reflect the assumption that the migration is basically a clockwise ellipse or triangle and that the birds are relatively route-faithful from year to year.

Ranges and the Sierra Nevada, and 2) along the Rocky Mountain Cordillera. Rufous nesting inland in southeast British Columbia, Alberta, Idaho, and Montana can fly down the Rockies via Colorado and New Mexico to the Sierra Madre Oriental, and we might expect all the breeders and offspring from coastal areas of Oregon, Washington, British Columbia, and Alaska to use the Coast Range–Sierra route.

From mid-July through August the Coast-Sierra migrants are refueling at 5,500 to 7,900 feet (1,700–2,400 m) in the Trinity Alps and Sierra Nevada of California, where good habitat gets packed to capacity. Those that miss out (usually females and juveniles) settle for suboptimal habitat. Especially for the urgent energy needs early and late in the day, they can attempt to supplement meager feeding by intruding into the better, already claimed territories. Patterns of habitat use are probably the same along the Rocky Mountain flyway.

Banding

Recaptures of banded birds are about as rare as direct observations (Plate 10.4). Catching and banding migrants is like eating Crackerjacks. We process many "sugared kernels" before finding the "prize" of a previously banded hummingbird. Recaptures from faraway places, too unlikely to justify research time and funding, are always cherished fantasies even though we catch the hummingbirds to get counts and body masses. In 12 years of netting at RMBL, only two of 4,139 rufous migrants came already banded from a different state. Two more, banded at RMBL, were subsequently reported in Washington and British Columbia. These recaptures of banded birds between the Rocky Mountains and the Pacific Northwest (see Map 10.1) suggest that many of the coastal birds fly inland to follow the Rocky Mountain flyway southeasterly (Table 10.1).

Only nine banded rufous have been recaptured across one or more state lines, two of which were recaptured more than once, for a total of 11 interstate recaptures. Only two of the nine birds were recaptured during a single migration (see Table 10.1 and Map 10.1). Five of these nine birds apparently were from the Pacific Northwest breeding distribution, but used the Rocky Mountain flyway to go south in the July–September migration.

Bob and Martha Sargent banded a rufous female in Pensacola, Florida, in December 1989 and recaptured her in nearby Bay Minette, Alabama, in February 1992. In November 1994, Sarah Driver found her in Springfield, Missouri, an amazingly unlikely series of chance encounters. Just where the Dixie winter residents breed we won't know without defying even more overwhelming odds.

Route fidelity was documented by recaptures of 39 birds that we had banded during a previous year's migration stop at RMBL. Such information has been slow in coming. Hummingbirds and their bands are obviously very small and unlikely to be noticed when life ends in nature. Consequently, recoveries of bands from corpses are very rare as well. In the vicinity of Gothic, Colorado, southbound rufous stop to refatten and rest over a period of 6 weeks in July and early August. Judging from the intensity of the avian bickering at flower patches and feeders, nobody is resting; it is strictly a matter of every bird for itself. Even though few of the banded birds show up elsewhere, the band numbers do provide a record that can be compared from year

Table 10.1 *Banding recaptures and recoveries of rufous hummingbirds*

Where banded[a]	Season banded[b]	Where recovered	Season recovered[b]	Elapsed time	Derivation of data
PNW–PNW					D. Bystrak,
x02212 California	N	Oregon	Br	21 mo	pers. comm.
RMt–RMt					W.A. Calder &
T12826 Montana	S	Colorado	S	15 d	E.G. Jones, 1989
PNW–RMt					G. Brown &
T78037 Vancouver Is. (BC)	Br	New Mexico	S	42 d	J. Day-Martin,
					pers. comm.
RMt–PNW					G. Brown,
T78037 New Mexico	S	Vancouver Is. (BC)	Br	10 mo	pers. comm.
RMt–PNW					S. Williamson,
T52675 Arizona	S	Washington	N	8 mo	pers. comm.
RMt–PNW					D. Bystrak &
T38968 Colorado	j/S	Washington	N/Br	9 mo	this study
RMt–PNW					D. Bystrak &
T31618 Colorado	S	Br. Columbia	Br	11 mo	this study
RMt–RMt					This study
T32623 New Mexico	S	Colorado	S	11 mo	
PNW–RMt					B. Wiard,
T37268 Alaska	S	Colorado	S	22 mo	pers. comm.
?–?					R. Sargent,
T16428 Florida	W	Alabama	W	26 mo	pers. comm.
?–?					R. Sargent,
T16428 Alabama	W	Missouri	S	31 mo	pers. comm.

Source: Updated from Reference 3.
[a]PNW = Pacific Northwest; RMt = Rocky Mountain area.
[b]N = northbound migration; Br = breeding; S = southbound migration; j = juvenile; W = winter.

to year, and within the season, to get the proportions of males vs. females and adults vs. juveniles for the following picture. The rusty-colored adult males, their gorgets like glowing coals of red-orange, are the first to migrate south via RMBL. They are followed by a few females, whose backs, necks, and crowns are iridescent green and whose throats show only spots of red-orange. Females gradually increase to numbers exceeding the males, lagging an average of 6 days behind them. Most adult rufous disappear from their breeding homeland ahead of the young. On their first migrations, the juveniles begin to appear after the peak of the adult passage. The midpoint in the flux of juveniles came 8 days after the midpoint in the female migration and lagged 2 weeks behind the passage of the adult males.

The instructions for the long migratory flights that lie ahead for recent fledglings must have been encoded in the egg for copy to the brain of the fledglings. This "hard-wired" migration pattern will be followed without parental guidance. One such juvenile rufous was banded at RMBL by Denise Stevenson on 6 August 1983, when most of the rufous adults had already passed through. On 18 July 1984, we recaptured him (now a spectacular adult male) 19 days earlier in the season, as was appropriate for his adult-male status. He visited our nets again on 7 July 1985. Over the years, seven birds banded as juveniles were recaptured at RMBL in at least one subsequent southbound migration.

Young birds that have never migrated south before must go it alone. Those that make it will be those that practice not only fidelity to birthplace/breeding ground and to their winter homes, but fidelity to a successful route, which is reflown in subsequent years, using the memory storage in 200 mg of brain. However, the birds are not automatons. The course and timing of migration is influenced by the seasonal climate. After unusually heavy winter snowpack and a late meltout in Colorado, we see fewer adult males at RMBL. They must be stopping at lower elevations where flowers are more abundant. Up at Gothic's 9,540-foot elevation, the flowers might not be ready and attractive for another week or two, when females and juveniles are arriving in numbers. On the other hand, if the winter has been mild and followed by an early melt and flowering, we capture proportionately more adult males.

Fueling for Migration

With captured birds briefly in hand, we record dates, numbers, mass, and condition (Plate 10.5). Such information is quite revealing. Seasonal shifts in feeding behavior and in the regulation of body mass represent profound changes in a hummingbird's physiology. During nesting, a female has little time for overeating; in territorial pursuits, males would want maximum acceleration from the force their pectoral muscles can develop. In contrast, the emphasis during the southbound migration is on maximum fat storage for the longest possible flights (Fig. 10.3). A rufous hummingbird can really pack on the fat. Preparing for a leg of the trip to Mexico for the winter, body weight can increase by two-thirds. The body mass of rufous migrants captured in Colorado and New Mexico ranged from 2.7 g (empty) to 5.7 g (full). Lynn Carpenter and colleagues found a similar range in southbound rufous stopovers in California. From 2.7 to 3.5 g, the weight gain was hypothesized to represent a rebuild-

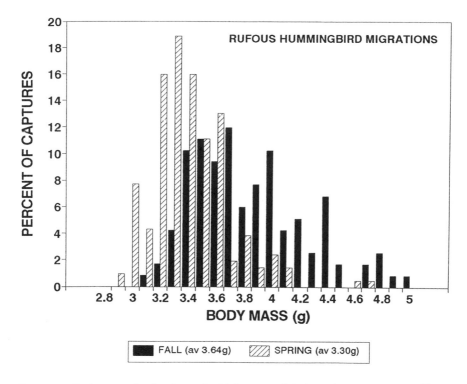

Figure 10.3 Body mass of rufous hummingbirds captured during migratory rest stops. New arrivals would be low (2.7–3.0 g). In the southbound migration, there are proportionately more of the heavier birds, which we may assume are closer to being refueled for the remainder of the flight.

ing of proteins burned in the migration segment just flown; above 3.5 g, the increase was entirely fat [11].

A hummingbird cannot convert all of the nectar's energy to fat. First, there are the costs of feeding activity. Hummingbirds must fight for access to flowers, hover while feeding, and then defend their flower patch from intruders. To maintain altitude, the force of lift must equal the downward force of the bird's weight. Hovering is the most expensive mode of flight, because there is no forward flight to generate part of the required lift; all the lift must come from wing movements in calm air.

Sixteen percent of the calories consumed must be spent to convert nectar sugar into stored fat. The added weight of this fat must then be lifted against gravity, resulting in shallower maximum-climb angles during flight after fueling. Climb angles of rufous hummingbirds when released after banding (Fig. 10.4) are affected by several factors, such as gender, lighting, and perception of the best escape path. The re-

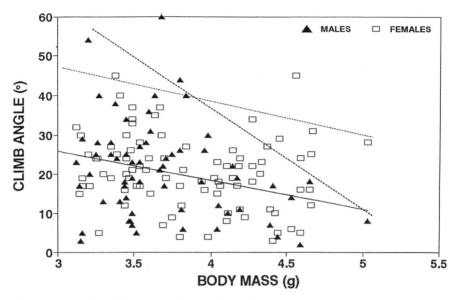

Figure 10.4 Angle of climb upon release of banded rufous hummingbirds (o = level flight). Self-fed with sucrose solution, the birds accumulated fat while fueling for continuation of their southbound migration. The solid line is the best fit for all the data (least = squares regression). The dotted lines show the approximate upper bounds for females (*light dots*) and males (*heavy dots*). Females have more wing surface area for lift and are less affected by added mass; hence the shallower slope to the female line.

sult is a range in observed climb angles—from level to as steep as possible—for birds of a particular weight. However, maximum climb angles were inversely related to bird mass. The heaviest of these self-fed birds were limited to a maximum climb angle of about 30°, while light birds could attain initial climb angles of 45° or more. Rufous males have shorter wings than the females and, thus, a smaller wing area for generating lift; so, with heavier *wing-loading* (the ratio of body mass to wing area), a male's climb is affected more by his fuel load.

Takeoff and climb-out are just the beginning of the next leg of the journey back into Mexico. At cruising altitude, more lift must be generated to keep a heavier bird flying. The cost per mile subsequently decreases as fat is burned, the bird's weight decreases, and less muscle is needed to fly. Even though protein contains only half as much energy per gram as fat, burning muscle protein allows the bird to fly farther southward before stopping to refuel.

Miles per gallon? How about Kalispell, Montana, to Yellowstone National Park on a gram of fat? Once fully-fattened, a hummingbird the size of a rufous could fly an estimated 610 miles (975 km) of its migration route without refueling and with no wind—that is to say, from Gothic, Colorado, to Silver City, New Mexico. By waiting for a tail wind after a frontal passage, the rufous can fly farther.

Winter

Many rufous hummingbirds spend part or all of the winter in the state of Jalisco, Mexico. Recaptures of 20 individuals at one to four years after banding in Parque Nacional Volcán Nevado and Sierra Manantlán Biosphere Reserve demonstrated their winter site-fidelity. Within-season recaptures of 12 banded birds at the same site suggest that much of the winter is spent in established residence, rather than in a progressive wandering throughout the winter range.

Northward Migration

The northbound, late-winter/spring rufous migration crosses northwestern Mexico into southwestern Arizona and California and then proceeds up the coast. Individual rufous reach northern Sonora and southern Arizona in mid- to late February, but significant numbers may not be encountered until a month or more later. Rufous arrive in Washington as early as 26 February–12 March, when stragglers are still 1,800 miles (3,000 km) south in Jalisco. Inland arrival of Rocky Mountain birds is later than at similar latitudes along the coastal states. Whereas some reach Juneau, Alaska, by 18 April, the mean first appearance (1968–1984) in Swan Valley, Montana, was 7 May [Ed Foss, pers. comm.].

Body mass at capture in spring migration ranges from 2.8 to 4.6 g (mean 3.27 g) versus 2.6 to 5.7 g (mean 3.44 g) when migrating southbound. This difference could be explained in two ways: 1) food is scarcer during the spring migration, so complete fattening is not feasible (*resource scarcity hypothesis*); or 2) the spring migration is flown with discretion, in shorter segments so as not to overshoot the leading edge of spring's emergence and the availability of nectar (*discretion hypothesis*).

The discretion hypothesis was supported in a study by Sara Hiebert, who maintained rufous hummingbirds for nearly a year in a controlled environment of 5°C at night and 20 °C in the daytime, with day length adjusted to correspond with the season outside. Even with unlimited food, their weights were lower in spring than at the time of southbound migration, similar to rufous in nature. Thus, the difference in maximum body mass between southbound and northbound migrations seems to have a programmed physiological basis [12].

It is stretching our anecdotal information to build a general picture of rufous hummingbird migrations from the bits and pieces that we have. One can only assume that others must be behaving similarly to the limited number of banded birds. But were these exemplars atypical, or were they representative of the species as a whole? Science may not be much help yet, but if there were no mysteries left to unravel, science would not be much fun. If we knew it all, we could not stand in awe of a naive, month-old bird with little more than $\frac{1}{10,000}$ of our brain volume bravely setting forth and successfully navigating to Mexico for the winter, returning north by an entirely different route, and then following that same loop the next year.

The Future of Rufous Migrations

According to the National Audubon Society's "Watchlist," the rufous hummingbird is one of the 14% of bird species breeding in the United States for which there are preliminary indications of decline. Censuses of small and highly mobile birds are crude, at best, so some apparent declining trends may need confirmation with more information. The annual reproductive potential of migrant hummingbird species is so modest that the only prudent course is to step up research and monitoring efforts and study the possible causes carefully. A true rufous decline could be explained by 1) loss or degradation of breeding habitat, 2) loss or degradation of winter habitat, and/or 3) migration casualties.

Breeding Habitat

Rufous hummingbirds are most abundant in secondary successional habitats, after forest fires, logging, and road construction. Considering the extent of deforestation in the Northwest and the Rockies, I have difficulty imagining that a decline in rufous is owing to a lack of breeding habitat.

Winter Habitat

Like several small passerines wintering in Mexico, the rufous hummingbird has broader habitat preferences in winter, occupying a range of elevations and latitudes where logging and burning are extensive. Mexico is a "hot spot" for speciation of tubular *Salvia* flowers, which the rufous seek and claim at sites that are generally at higher elevations than areas where there is intensive use of agricultural pesticide. More work on winter biology is needed, but there is as yet no evidence that winter is the hummingbirds' Achilles' heel.

Hazards during Migration

Gass and Lertzman have described the impact of a violent thunderstorm that hit a subalpine meadow in the Salmon-Trinity Alps, California, destroying most of the red columbine flowers on which migrants had been feeding. Within hours, this resulted in an 88% reduction in the number of territories of transient rufous, and that number remained low for the rest of the migratory season. Presumably the former occupants were forced into suboptimal habitats [13]. The late Alden Miller told of the death of an underweight (2.5 g) rufous after it failed "during a migratory flight in hot sun through an area without water or nectar sources" (Joshua Tree National Monument in the Mojave Desert) [14].

Over broad regions, the timing and amount of precipitation must often determine the success of migratory nectar-feeders. In the White Mountains of eastern Arizona, winter precipitation was far below average between 1974 and 1977, leaving a scant snow cover that melted early in the spring. The result was summer drought, heavy mortality in the flower species used by hummingbirds, and intense competition for the low-nectar standing crop: "many birds emigrated, but many others undoubtedly died of starvation" [15] (Plate 10.6). Rufous hummingbirds are clearly vulnerable to weather-related variation in flower nectar availability. In 7 summers of study at a stopover site in sagebrush shrub desert at the base of the eastern Sierra Nevadas, R.W. Russell and co-workers found that a higher percentage of the birds weighed less than 3 g when flowers, especially Indian paintbrush, were scarce. The average stopover duration of marked individuals was longer, and hummingbird traffic was least in drought years, when flowers were scarcer [16].

In spring my colleagues and I mist-net and band birds along a north–south arroyo at the western base of Cerro Suvuk in the Sierra Pinacate Biosphere Reserve in northwestern Sonora, Mexico. Following a winter drought in 1995–1996, most of the chuparosa had withered in bud and there were hardly any birds. Two visits (12–13 and 22–24 March) netted only three very light female Costa's hummingbirds (local breeders and migrants). Two male and one female rufous were present, but avoided the net. One male rufous defended one of the only chuparosa clumps having a few flowers, but when no other hummingbird was around, he perched low, often almost lying on a dirt bank in the morning, his eyes nearly closed. I crept up and moved my hand to within a foot of him before he flew. He was still alert but did not move unless necessary. I assumed he was saving what little energy he had, in a probably vain hope of reaching the Northwest to breed. By 22 March, chuparosa flowers and rufous hummingbirds were absent. Just two female and one male Costa's fed on ocotillo.

Up in the saddle between Carnegie and Pinacate peaks, a condensation fog or cloud belt of moisture from the adjacent Gulf of California usually sustains a brilliant stand of magenta-red Arizona *Penstemon* (bearded tongues) in March, attracting many migrant rufous hummingbirds. However, in 1996 we saw only six *Penstemon* flower stalks on the entire south side of Carnegie and the adjacent saddle. Dry *Penstemon* stalks from spring 1995 were abundant, so we were in the right place, but in the wrong year. Such a drought was repeated in 1997. Two consecutive years is long enough for a generation of hummingbirds to have died, and such sustained climatic vicissitudes, if widespread, could have a significant impact on the species' population.

How do migrants pass through on drought years? Can they find alternative routes? Will drought years become more common with climate change? Is drought-caused flower failure the reason for apparent decreases in rufous hummingbird populations? We simply don't know. Habitat protection is important, but the greatest vulnerability may come from the effects of climate change. Perhaps rapid evolution of changes in migratory behavior can provide the flexibility needed to cope with climate change, as suggested by the exponential increase in the numbers of rufous wintering in the southeastern United States. Between 1909 and 1979, rufous hummingbirds reported from the southeastern states numbered 68. Between 1990 and 1994, the number was 869. However, we should not count on that.

The consequences for the smallest birds in natural drought years may have a message: Could declines in the abundance of Neotropical migratory birds be a consequence of our effects on climate as well as on more direct factors such as habitat loss? We live in a time of exponential growth in human demand for natural resources and of ever-increasing impacts on ecosystem function. The ramifications of these impacts are poorly understood, and the willingness of politicians to allocate funds for environmental protection and research has declined just when the need for information is most crucial. The appeal of birds to nonbiologists has always been a great asset for biology, and now our birder colleagues can be of special service—censusing, studying species or areas on a long-term basis, and banding to detect the trends, good or bad.

References

1. Grant, K.A., and V. Grant. 1967. Hummingbirds and Their Flowers. New York: Columbia Univ. Pr.
   A popular treatment of hummingbirds and hummingbird flowers.
2. Johnsgard, P.A. 1997. The Hummingbirds of North America (2nd ed.). Washington, DC: Smithsonian Inst. Pr.
   A general account of hummingbird life histories.
3. Calder, W.A. 1993. Rufous hummingbird. *In* The Birds of North America, No. 53 (A. Poole

and P. Stettenheim, eds.). Washington, DC, and Philadelphia: Am. Ornithol. Union and Acad. Nat. Sci.

The most extensive life history of the species.

4. Calder, W.A. 1987. Southbound through Colorado: migration of rufous hummingbirds. Nat. Geog. Res. 3:40–51.

A popular account of rufous hummingbird migration in Colorado.

5. Calder, W.A. 1991. Broad-tailed hummingbird. *In* The Birds of North America, No. 16 (A. Poole, P. Stettenheim, and F. Gill, eds.). Washington, DC, and Philadelphia: Am. Ornithol. Union and Acad. Nat. Sci.

The most recent account of the life history.

6. Ehrlich, P.R., and G.C. Daily. 1988. Red-naped sapsuckers feeding at willows: possible keystone herbivores. Amer. Birds 42:357–365.

An interesting description about how the work of sapsuckers makes resources available to other species, including hummingbirds.

7. Cobb, S. 1966. The brain of the emu *Dromaeus novaehollandiae:* II. Anatomy of the principal nerve cell ganglia and tracts. Brevoria 250:1–16.

A technical anatomical treatise.

8. Willimont, L., S. Senner, and L. Goodrich. 1988. Fall migration of ruby-throated hummingbirds in the northeastern United States. Wilson Bull. 100:482–488.

A study of migration during daytime by direct visual observation.

9. Carpenter, F.L., D.C. Paton, and M.A. Hixon. 1983. Weight gain and adjustment of feeding territory size in migrant hummingbirds. Proc. Natl. Acad. Sci. U.S.A. 80:7259–7263.

A technical paper, now a classic, that demonstrates the trade-off between the costs of territory defense and the food reward obtained from that territory.

10. Phillips, A.R. 1975. The migrations of Allen's and other hummingbirds. Condor 77:196–205.

A description of migration patterns; based on distributional information.

11. Carpenter, F.L., C.A. Beuchat, R.W. Russell, M.A. Hixon, and D.C. Paton. 1993. Biphasic mass gain in migrant rufous hummingbirds: body composition changes, uses of torpor and ecological significance. Ecology 74:1173–1182.

A technical paper examining various interactions in the energetics of hummingbirds during stopover.

12. Hiebert, S.M. 1993. Seasonal changes in body mass and use of torpor in a migratory hummingbird. Auk 110:727–797.

A technical paper discussing torpor in hummingbirds during the annual cycle.

13. Gass, C.L., and K.P. Lertzman. 1980. Capricious mountain weather: a driving variable in hummingbird territorial dynamics. Can. J. Zool. 58:1964–1968.

A technical paper dealing with the costs and benefits of territoriality.

14. Miller, A.H. 1963. Desert adaptations in birds. Proc. XIII Intern. Ornithol. Congr., pp. 666–674.

Discusses the rigors of desert living in a variety of birds, including the rufous hummingbird.

15. Brown, J.H., and A. Kodric-Brown. 1979. Convergence, competition, and mimicry in a temperate community of hummingbird-pollinated flowers. Ecology 60:1022–1035.

A technical paper analyzing the dynamics of the whole ecological community to which hummingbirds belong in southeastern Arizona.

16. Russell, R.W., F.L. Carpenter, M.A. Hixon, and D.C. Paton. 1994. The impact of variation in stopover habitat quality on migrant rufous hummingbirds. Conservation Biology 8:483–490.

The effects of drought on the use of stopover habitat and the energetics of migrant rufous hummingbirds.

Epilogue KENNETH P. ABLE

Conservation
of Birds
on Migration

An endangered phenomenon is a spectacular aspect of the life history of an animal or plant species involving a large number of individuals that is threatened with impoverishment or demise; the species per se need not be in peril, rather the phenomenon it exhibits is at stake. —Lincoln P. Brower and Stephen B. Malcolm [1]

If you have read the foregoing chapters of this book, I cannot imagine that you will not have been impressed by the immensity, elegance, and inherent risk that attend the great migrations we have discussed. It is easy to be deceived into thinking about them as a collective — great swarms of migrants flowing this way and that, a population phenomenon. But the migrant flocks of shorebirds and cranes and the broad-front tide of songbirds are made up of individual birds, each using whatever skills and tricks it can muster to endure the flights, weather the storms, find safe havens in which to rest and refuel; in short, to carry on, survive, and reproduce. It is at the level of individual survival and reproduction that the health of our migrant populations is played out.

We deliberately tried to avoid writing a "gloom and doom" polemic about the future of migratory birds. It seemed to us that any intelligent person, informed about the magnitude of the feats performed by these migrants and the exquisite adaptations with which they perform them, would perforce conclude that preserving these birds, their populations and, indeed, the migration syndrome itself, is an extraordinarily worthy goal.

Anyone with even a casual interest in birds cannot have failed to notice the barrage of press coverage devoted to the reported declines in and conservation of migratory birds. John Terborgh raised an alarm nearly a decade ago with his book

Where Have All the Birds Gone? [2]. Since then, both the conservation literature and the popular press have been replete with reports of the decline in populations of migratory birds, especially songbirds that migrate into the tropics for the winter. Some of the reporting has been wildly hyperbolic. In its Winter 1990 issue, the magazine *American Birds* published an article with this screaming title: "2001: Birds That Won't Be with Us." Among other species, it listed the golden-winged warbler, whose portrait was emblazoned on the cover. If any of the species listed is to be gone by the millennium—with the exception of Bachman's warbler, which was extinct, or nearly extinct when the article was written—it had better hurry up. It is easy to get the impression from what one reads in much of the popular press that you may have seen your last wood thrush.

Many of us who care deeply about conservation in general, and about the preservation of migratory birds in particular, are concerned that such exaggerated rhetoric has the potential to erode the very support and commitment that will need to be brought to bear if we are to conserve these migrants over the long term. The issues are fundamentally scientific ones, but the solutions to the problems will require public support and political action. It is thus critical to keep a clear view of what we know about migratory bird populations, the degree of certainty with which we know it, and the inferences that can confidently be made from those data.

As one might expect, there are significant differences of opinion among the experts as to the extent and especially the proximate causes of declines in migratory bird populations [3]. It is frustratingly difficult to be certain that the populations of relatively numerous and widespread species are really suffering long-term declines. As an illustration, imagine that it were possible to know the trends in the populations (i.e., increasing, decreasing, remaining the same) of 40 species of birds at a given instant in time. Even if every one of those populations was in *long-term* equilibrium, neither increasing nor decreasing over the long haul, we would expect just by chance alone that some would be increasing, some would be decreasing, and some would not be changing in our instantaneous snapshot. Even if we could census populations with precision, itself a very thorny problem, we need standardized, statistically rigorous long-term census data to be certain about long-term population trends. Very few such data sets exist. In North America, the Breeding Bird Survey, begun in 1966, provides the best census tool, but most of the boreal forests of Canada where many of our migrants nest (e.g., the spruce-woods warblers) are beyond the reach of its roadside coverage. In addition, the methods used to analyze the data and the conclusions drawn remain the subject of lively debate among biologists [3]. So what do we know?

Scott Robinson, whose work on the ecology of migratory birds spans the Neotropics to the forests of the Midwest, recently wrote a very balanced summary of the state of our knowledge about migratory songbird populations [4]. Much of the popular press reportage on his article notwithstanding, the main conclusions of Robinson's analysis were as follows: 1) *taken as a whole* and *averaged over large geographic areas,* Neotropical migrant populations have remained generally stable over the 30 years of the Breeding Bird Survey; 2) some forest-dwelling species such as the wood thrush and cerulean warbler are declining steeply even in the core areas of their breeding ranges (at average rates of 1% to 3% per year over the past 30 years); 3) in some regions, many or even most forest passerines are declining; 4) some species of shrubland birds have declined as rapidly or even more rapidly than forest-dwellers; and 5) grassland birds have shown the largest and most consistent downward trends.

Robinson's synthesis provides one credible interpretation of what the best available data tell us. It is not the rosy picture painted in some of the newspaper coverage of his paper, nor is it the doomsday scenario advocated by some in the conservation movement. Robinson's analysis of the situation is also not exactly concordant with the impressions of many long-time field birders. Veteran watchers along the Gulf Coast are adamant that the frequency and magnitude of fallouts today are only a shadow of what they were 30 years ago. Birders in New England have witnessed season after season of lackluster migrations: the hundreds of blackpoll warblers that Jim Baird used to see just in his yard during fall migration have become single-digit counts. Although these sorts of observations and others (e.g., capture rates at long-standing banding stations) are backed up by numbers, they are often unsystematic and necessarily local or, at best, regional in scope. And yet the same message has been coming from many quarters for more than a decade now. Are we being fooled by failing senses or faulty memories? I am inclined to think not. Whereas the sky may not be falling, sustained population decreases of even 1% or 2% per year are intolerable over relevant ecological time scales: such trends will lead inexorably to the disappearance of these species. Is a 30-year record sufficient to predict what will happen over the next century? No one knows, but the Breeding Bird Survey model is the best predictive tool we have. On the basis of what it tells us, I think it is surely prudent to take warning now, to attempt to find out what is going on and implement plans to ensure the long-term stability of these bird populations before they become threatened or endangered and before the problems and cost of saving them becomes insurmountable.

Being able to document convincingly a long-term downward trajectory in a bird population is hard enough, but that is only a meager beginning. If we wish to invoke conservation measures, we must in addition discover *why* the population is decreasing. Here the problem is compounded. In discussions concerning the causes of population declines among migratory birds, most of the focus has been on events and conditions on the wintering and breeding grounds, especially in the case of songbirds: e.g., habitat fragmentation and the attendant increase in predation and cow-bird parasitism in the North, tropical forest destruction in the South. Yet many migratory birds spend more of their lives moving between these two places than they do in either one and, as I hope this book has amply demonstrated, the challenges to survival may often be even greater during migration.

Two themes transcend nearly all the chapters of this book. One is that long-distance migration is an inherently risky and stressful endeavor. The second is that this endeavor is played out on a global stage. The necessary habitat for migratory birds is a string of islands linked together by the birds' movements.

Some anthropogenic change that is likely to have an impact upon migratory birds is occurring at a relatively slow rate (though still much faster than most natural environmental change). Global warming, for example, is certain to affect migration patterns in a variety of ways. Peter Berthold [5] has pointed out that, among other things, global warming at higher latitudes is likely to favor resident species (e.g., by leading to a reduction in winter mortality). Such a change could have a negative competitive impact upon migrants. But most of the man-induced environmental changes that may impact migratory birds are taking place at a much more accelerated pace.

Is it reasonable to think that migratory behavior can change rapidly enough to enable species to cope with these kinds of changes? In Chapter 1, I described the evolution of new migration patterns in European blackcaps and North American house finches, change that has occurred on the order of half a century. In Chapter 10, Bill Calder notes the new pattern of overwintering by rufous hummingbirds in the southeastern states. So hope exists that populations may in some cases be able to adapt fairly rapidly. The speed with which they can do so will depend in part on the extent of heritable variability in migratory traits that exists within the population. As Berthold notes, partial migrants are likely to be at an advantage in this situation because their populations already contain migrants and nonmigrants: there is considerable behavioral variability within the population upon which natural selection can act (assuming that the behavioral variability has a genetic basis). Variability is also generally a function of population size, so maintaining large populations of migratory species is good insurance. Obligate long-distance migrants with more constrained behavior and physiology are likely to be at greater risk. As noted in Chapters 1 and 2, the evolution of migratory behavior involves the acquisition of an entire suite of interrelated adaptations: it is a syndrome. Obligate migrants have gone farther down the path of specialization, and thus their evolutionary options over the short term may be more limited.

The stories that unfold in the chapters of this book reinforce the general consensus among biologists that the main threat to migratory birds (and other animals as well) is habitat loss. This is why birds of grasslands and wet meadows have sustained the most dramatic recent population declines. Most populations can survive a catastrophic event and recover former numbers so long as the habitat they require remains intact. This is not news, but it is easy to become diverted to peripheral issues and lose sight of the fact that, at the end of the day, we will not preserve migratory birds if we do not protect adequate quantities of the necessary habitats. Concerns about the effects of climate change will be irrelevant. The necessary habitats include not just the breeding and overwintering areas, but all the stopover islands in between. The mobile life-style of migrants makes conservation problems involving migratory birds both more threatening to the birds and confounding to those who would attempt to solve them. The vital islands for a given species may include quite different habitats (think of the white-rumped sandpiper utilizing coastal mudflats and inland freshwater habitats) and may span continents and many different political entities. Any one of these islands, if jeopardized, can become the migrant's "Achilles' heel" [6]. Yet, even identifying all of the critical places, much less protecting them, is a daunting challenge. Some migrants congregate in vast numbers at a very small number of places. Shorebirds are notable for this habit, as Stan Senner described with respect to the Copper River Delta. The spring concentration of red knots feeding on horseshoe crab eggs in the Delaware Bay or the staging of white-rumped sandpipers and other shorebirds at Cheyenne Bottoms are other well-publicized examples. In some of these cases, a very large proportion of a continental population of a species may be concentrated simultaneously in one small area. On one hand, this sort of concentration can make conservation measures easier because it is obvious which site or sites must be protected. On the other hand, such aggregation makes the birds far more vulnerable to a local disaster, whether man-made or natural. That sort of vulnerability was recently illustrated in the Delaware Bay, where threats to migratory shorebirds resulting from an overharvest of horseshoe crabs resulted in a moratorium on human harvesting of the crabs.

For songbirds, it is harder to know where the most important places are. Places to which birders flock to see migrants are not always places where the migrants themselves assemble in the greatest numbers, nor necessarily are they particularly favorable stopover sites from the birds' perspective. Islands such as the Dry Tortugas in Florida or the equally insular greenery provided by small urban parks may often attract spectacular concentrations of grounded songbird migrants. Superficially, such places would appear to be important habitats for en route migrants. However, many such places offer inadequate food, water, and other necessities for the passage migrant and may turn out to be more a Siren's trap than an oasis. On the Tortugas it is a common but disturbing sight to watch migrant cattle egrets slowly weaken and die because there is virtually nothing to eat (Plate E.1). Many might not have been able to make it to the mainland anyway, but those with any chance should have kept on going. In evaluating the importance of stopover sites, then, we need to know whether the birds that land there can recover, refuel, and embark in a timely manner on the next leg of their journey. That is much more difficult to determine than the numbers of migrants that stop.

The new network of WSR-88D weather radars that Sid Gauthreaux discussed in Chapter 3 provides a potentially important tool that may better enable us to identify important stopover areas for migratory songbirds on a regional scale. Just after dark, as the migrants initiate their night flights, the radar will reveal where the densest concentrations are taking off. We observed this pattern many years ago with the previous generation of weather radars. In coastal southwestern Louisiana, where oak cheniers and small woodlands are surrounded by a sea of marsh and grassland, the pattern of birds on the radar just after dark revealed quite precisely where the wooded areas in which they had spent the day were located.

There seems to be little doubt that the coastal woodlands provide critically important habitat for trans-Gulf migrants, especially birds arriving in spring that are either low on fat or dehydrated. Being able to stop there may often make the difference between life and death. But recent studies with the new radar have shown that the largest numbers of spring trans-Gulf migrants emerge neither from the coastal areas nor from the drier pine forests, but from the wet bottomland riparian forests. These areas are difficult of access and characterized by dense vegetation, tall trees, and an abundance of biting insects and venomous reptiles. Determining their importance for migrants by ground-based surveys would be nearly impossible. The radar provides an instantaneous, broad-scale, and quantitative overview (Plate 11.2).

Identifying the most important habitats and sites used by birds on migration is a necessary first step. Protecting these places is the next and more difficult step. Over the past few years, migratory birds have attained high visibility: they serve as a flagship group of animals, symbolizing broader concerns about the loss of biodiversity. Organizations such as the National Fish and Wildlife Foundation's "Partners in Flight," the Nature Conservancy, and the National Audubon Society have launched programs specifically targeting the conservation of migratory birds, especially Neotropical migrants. Federal and state agencies responsible for the management of public lands are now mandated to consider the welfare of migratory birds in their planning processes. Legislation is pending at the federal level that would tax outdoor recreational equipment such as binoculars, with the proceeds flowing to state nongame wildlife conservation programs.

Many local efforts are also under way, including those on the northern Gulf Coast that resulted in the protection of the Peveto Beach Woods in Louisiana, where Frank Moore works (Chapter 4), and the improvement of habitat for migrants on Dauphin

Island, Alabama (Fig. 11.1). Increasingly, local communities in places where conspicuous migration events occur annually are attempting to cash in on that fact by hosting various sorts of birding festivals. In well-known places such as Cape May, New Jersey, the revenue generated by this "new" tourism can be a significant contribution to the local economy and a stimulus to protect the habitat that draws the birds. The number of people actively interested in birds as a hobby is increasing dramatically, and birders spend billions of dollars each year on the goods and services required to watch and feed birds [7]. They represent a potentially very powerful constituency to press government agencies and other groups for effective action to protect migratory birds. The resulting political clout has the potential to bring about habitat preservation.

All of these sorts of endeavors are commendable and, to the extent that they result in protection or restoration of important habitat used by migrants, are worthwhile. However, we need effective means to evaluate what our conservation programs are actually accomplishing in terms of benefits to the birds on the ground. Media events, meetings, attractive articles, and priority lists may all be necessary components of the process, but if they don't result in preservation or restoration of vital habitat or greater protection for migrants, the efforts are hollow at best, cynical at worst. How are we doing? Certainly the issue of migratory bird conservation has been brought to the forefront. It is on the agenda of governmental agencies and nonprofit conservation organizations. Some innovative initiatives, such as advocating shade-grown coffee, are entering the mix. Still, over the growing number of decades during which I have been paying attention to migratory birds, the trends are not encouraging. I see more habitat being lost to five-acre house lots, to shopping malls, to road building, to cancerous and ill-advised coastal development, to clear-cutting. I see ever more habitat degraded by fragmentation, overgrazing, poor forestry and agricultural practices—and, for grassland and shrubland birds, by the inexorable ticking of plant succession. And despite some progress in cleaning up toxins in the environment, pervasive pollutants still appear to be having insidious effects on birds and other wildlife [8].

Whatever the particular immediate peril faced by a migratory bird, all of the man-made threats are either directly caused by or exacerbated by the unrestrained growth of the human population. Unfortunately, this is just as true in the less developed parts of the world as it is in the industrialized countries. In the former, direct pressure on the landscape for the necessities of survival destroys habitat, whereas in the latter an excessive per capita demand for resources contributes to pollution at home and habitat destruction both at home and abroad. Unless something dramatic can be done to curb the explosion of our own population, I find it difficult to be anything but pessimistic about the long-term future of migratory birds or about biodiversity

Figure E.1 Two views of the Shell Mounds woods on Dauphin Island, Alabama, showing the greatly improved habitat for migrants at this popular stopover site for trans-Gulf migrants. (*Top*) April 1969. (*Bottom*) April 1992. (Photos by K.P. Able)

in general. Most conservation organizations and nearly all governments seem loathe to confront this issue head-on; but it is the core problem, and if we cannot solve it the remainder of the environmental agenda will become moot.

I won't be around to see the diminished world that will result if we fail to preserve any significant amount of life's diversity. So why do I care? This is a philosophical issue, an arena in which I cannot claim any competence and with which I am not particularly comfortable. I am a scientist, and what science tells me is that for the first time since the very early days of life on earth, a single species has the capacity to change the entire future course of the planet and its life forms. Through a simple twist of evolutionary fate, we wound up being the species which, through the technological reach of our intellect, acquired that power. It seems to me that such power carries with it a moral responsibility to protect other creatures and their world—for their direct and indirect economic value, for the potentially useful genetic information they contain, for the functional roles they perform in healthy, stable ecosystems. All good reasons in themselves. In some cases, one can even estimate the economic value of the animals and services they perform in ecosystems. Such economic arguments can be used to support the conservation of biodiversity and habitat preservation, but they carry a risk. As soon as some competing interest claiming greater dollar value appears, the marsh is drained, the forest cut. It seems to me that to rely too heavily on strictly economic arguments for the conservation of biodiversity risks dooming us to the bean counter's dilemma—knowing the cost of everything but the value of nothing. Just as important, for me at least, is that without these birds, the world would be a very bleak place. To imagine a world in which the warblers, thrushes, and tanagers do not return to our forests in spring, where clouds of migratory shorebirds do not wheel and dart across coastal mud flats, or in which the great clamor of cranes cannot be heard in the heartland of North America is intolerably sad. Surely we want our children and theirs to be able to experience the natural wonders that we can still enjoy. Not just because they have economic value or provide a compound that can cure a disease, but because they are a big part of what makes life worth living and because it is the right thing to do.

If you have read this book, then you are very likely a member of the constituency that has a heartfelt interest in the future of these migrating "angels." The foregoing chapters have been intended to help you understand the problems and challenges faced by these migratory travelers. We hope that this information will better equip you to recognize effective conservation measures and motivate you to take constructive action on their behalf.

References

1. Brower, L.P., and S.B. Malcolm. 1991. Animal migrations: endangered phenomena. Am. Zool. 31:265–276.

   A readable article devoted primarily to a discussion of monarch butterfly migration.

2. Terborgh, J. 1989. Where Have All the Birds Gone? Essays on the Biology and Conservation of Birds That Migrate to the American Tropics. Princeton (NJ): Princeton Univ. Pr.

   The classic call to arms with regard to Neotropical migrants.

3. Maurer, B.A., and M.-A. Villard. 1996. Continental scale ecology and Neotropical migratory birds: how to detect declines amid the noise. Ecology 77:1–2.

   The introduction to a group of technical papers evaluating methods for monitoring migratory bird populations.

4. Robinson, S.K. 1997. The case of the missing songbirds. Consequences 3:3–15.

   A thorough, readable review and synthesis of the status and conservation of migratory songbird populations as indicated by the North American Breeding Bird Survey.

5. Berthold, P. 1993. Bird Migration: A General Survey. New York: Oxford Univ. Pr.

6. Myers, J.P., R.I.G. Morrison, P.Z. Antas, B.A. Harrington, T.E. Lovejoy, M. Sallaberry, S.E. Senner, and A. Tarak. 1987. Conservation strategies for migratory species. Am. Sci. 75:19–26.

   Focuses mainly on shorebirds.

7. [Anonymous]. 1997. What's a bird worth? Bird Cons., Spring Migration:6–8.

   A news account containing a variety of statistics about birding economics.

8. Colborn, T., D. Dumanoski, and J.P. Myers. 1996. Our Stolen Future. New York: Dutton.

   An important and frightening book about the potential of a wide range of common environmental pollutants to disrupt endocrine function in animals, with disastrous effects; numerous examples from bird studies are presented.

Common and Scientific Names of Birds and Plants Mentioned in the Text

Birds

American golden-plover *Pluvialis dominicus*

American kestrel *Falco sparverius*

American redstart *Setophaga ruticilla*

American robin *Turdus migratorius*

Arctic tern *Sterna paradisaea*

Bachman's warbler *Vermivora bachmanii*

Baird's sandpiper *Calidris bairdii*

Bald eagle *Haliaeetus leucocephalus*

Bar-headed goose *Anser indicus*

Black turnstone *Arenaria melanocephala*

Black vulture *Coragyps atratus*

Black and white warbler *Mniotilta varia*

Black-bellied plover *Pluvialis squatarola*

Black-throated blue warbler *Dendroica caerulescens*

Black-whiskered vireo *Vireo altiloquus*

Blackcap *Sylvia atricapilla*

Blackpoll warbler *Dendroica striata*

Blue tit *Parus caerulescens*

Bobolink *Dolichonyx oryzivorus*

Bristle-thighed curlew *Numenius tahitiensis*

Broad-tailed hummingbird *Selasphorus platycercus*

Broad-winged hawk *Buteo platypterus*

Calliope hummingbird *Stellula calliope*

Canada goose *Branta canadensis*

Cape May warbler *Dendroica tigrina*

Cattle egret *Bubulcus ibis*

Cerulean warbler *Dendroica cerulea*

Cliff swallow *Hirundo pyrrhonota*

Common snipe *Gallinago gallinago*

Common swift *Apus apus*

Common yellowthroat *Geothlypis trichas*

Connecticut warbler *Oporornis agilis*

Cooper's hawk *Accipiter cooperii*

Costa's hummingbird *Calypte costae*

Dunlin *Calidris alpina*

Eastern kingbird *Tyrannus tyrannus*

Emu *Dromaius novaehollandiae*

Eskimo curlew *Numenius borealis*

European robin *Erithacus rubecula*

European starling *Sturnus vulgaris*

Golden eagle *Aquila chrysaetos*

Gray-cheeked thrush *Catharus minimus*

Great blue heron *Ardea herodias*

Great kiskadee *Pitangus sulphuratus*

Greater white-fronted goose *Anser albifrons*

Greater yellowlegs *Tringa melanoleuca*

Green violet-ear *Colibri thalassinus*

Homing pigeon *Columba livia*

Hooded warbler *Wilsonia citrina*

House finch *Carpodacus mexicanus*

Indigo bunting *Passerina cyanea*

Kentucky warbler *Oporornis formosus*

Laysan albatross *Diomedea immutabilis*

Lazuli bunting *Passerina amoena*

Least sandpiper *Calidris minutilla*

Least tern *Sterna antillarum*

Lesser yellowlegs *Tringa flavipes*

Magnolia warbler *Dendroica magnolia*

Manx shearwater *Puffinus puffinus*

Merlin *Falco columbarius*

Mississippi kite *Ictinia mississippiensis*

Mourning warbler *Oporornis philadelphia*

Northern cardinal *Cardinalis cardinalis*

Northern goshawk *Accipiter gentilis*

Northern harrier *Circus cyaneus*

Northern mockingbird *Mimus polyglottos*

Northern pintail *Anas acuta*

Orange-crowned warbler *Vermivora celata*

Orchard oriole *Icterus spurius*

Osprey *Pandion haliaetus*

Ovenbird *Seiurus aurocapillus*

Pacific golden-plover *Pluvialis fulva*

Painted bunting *Passerina ciris*

Passenger pigeon *Ectopistes migratorius*

Peregrine falcon *Falco peregrinus*

Pine grosbeak *Pinicola enucleator*

Pine siskin *Carduelis pinus*

Piping plover *Charadrius melodus*

Purple gallinule *Porphyrula martinica*

Red knot *Calidris canutus*

Red-eyed vireo *Vireo olivaceus*

Red-necked phalarope *Phalaropus lobatus*

Red-shouldered hawk *Buteo lineatus*

Red-tailed hawk *Buteo jamaicensis*

Ross's goose *Chen rossii*

Rough-legged hawk *Buteo lagopus*

Ruby-throated hummingbird *Archilochus colubris*

Rufous hummingbird *Selasphorus rufus*

Sanderling *Calidris alba*

Sandhill crane *Grus canadensis*

Savannah sparrow *Passerculus sandwichensis*

Scarlet tanager *Piranga olivacea*

Semipalmated sandpiper *Calidris pusilla*

Sharp-shinned hawk *Accipiter striatus*

Short billed dowitcher *Limnodromus griseus*

Snow goose *Chen caerulescens*

Snowy owl *Nyctea scandiaca*

Summer tanager *Piranga rubra*

Surfbird *Aphriza virgata*

Swainson's hawk *Buteo swainsoni*

Tennessee warbler *Vermivora peregrina*

Tundra swan *Cygnus columbianus*

Turkey vulture *Cathartes aura*

Varied thrush *Ixoreus naevius*

Western kingbird *Tyrannus verticalis*

Western sandpiper *Calidris mauri*

Whimbel *Numenius phaeopus*

White-crowned sparrow *Zonotrichia leucophrys*

White-eyed vireo *Vireo griseus*

White-rumped sandpiper *Calidris fuscicollis*

Whooping crane *Grus americana*

Willet *Catoptrophorus semipalmatus*

Wood thrush *Hylocichla mustelina*

Yellow warbler *Dendroica petechia*

Plants

Barbey's larkspur *Delphinium barbeyi*

Bearded tongues *Penstemon* sp.

Chuparosa *Justicia californiana*

Corydalis *Corydalis caseana*

Currants *Ribes* sp.

fireweed *Epilobium angustifolia*

Fool's huckleberry *Menziesia ferruginea*

Grape vine *Vitis cinerea*

Greenbrier *Smilax tamnoides*

Hackberry *Celtus laevigata*

Honey locust *Gleditsia triacanthos*

Honeysuckle *Lonicera* sp.

Huckleberry *Vaccinium* sp.

Indian paintbrush *Castilleja* sp.

Live oak *Quercus virginiana*

Nelson's larkspur *Delphinium nelsoni*

Ocotillo *Fougquieria splendens*

Palmetto *Sabal minor*

Poison ivy *Toxicodendron radicans*

Quaking aspen *Populus tremuloides*

Ragwort *Senecio*

Red columbines *Aquilegia*

Red mulberry *Morus rubra*

Sages *Salvia*

Salmonberry *Rubus spectabilis*

Scarlet gilia *Ipomopsis aggregata*

Sitka spruce *Picea sitchensis*

Sweet acacia *Acacia farnesiana*

Toothache-tree *Zanthoxylum clavaherculis*

Wild tobacco *Nicotiana* sp.

Yaupon *Ilex vomitoria*

Contributors

KENNETH P. ABLE is Professor of Biological Sciences at the University at Albany, State University of New York.

JAMES BAIRD is recently retired from the Massachusetts Audubon Society, where he served as Director of its Conservation Department and, most recently, as Vice President for Special Projects.

KEITH L. BILDSTEIN is Director of Research and Education for the Hawk Mountain Sanctuary Association, Kempton, Pennsylvania.

WILLIAM A. CALDER is a Professor in the Department of Ecology and Evolutionary Biology at the University of Arizona, Tucson.

SIDNEY A. GAUTHREAUX JR. is Professor of Zoology at Clemson University, Clemson, South Carolina.

BRIAN A. HARRINGTON is Staff Biologist with the Manomet Center for Conservation Sciences, Manomet, Massachusetts.

GARY L. KRAPU is a Research Biologist with the Northern Prairie Wildlife Research Center, U.S. Geological Survey, Jamestown, North Dakota.

FRANK R. MOORE is Professor of Biological Sciences at the University of Southern Mississippi, Hattiesburg, Mississippi.

STANLEY E. SENNER is Science Coordinator, Exxon Valdez Oil Spill Trustee Council, Anchorage, Alaska.

Index